Universität Stuttgart

Institut für Energieübertragung und Hochspannungstechnik, Band 48

Objective Interpretation of Frequency Response Analysis of Power Transformers

Objective Interpretation of Frequency Response Analysis of Power Transformers

Von der Fakultät
Informatik, Elektrotechnik und Informationstechnik
der Universität Stuttgart
zur Erlangung der Würde eines Doktor-Ingenieurs (Dr.-Ing.)
genehmigte Abhandlung

vorgelegt von

Mehran Tahir

aus Gujranwala, Pakistan

Hauptberichter: Prof. Dr.-Ing. Stefan Tenbohlen
Mitberichter: Prof. Dr.-Ing. Ebrahim Rahimpour
Tag der mündlichen Prüfung: 25.11.2024

Institut für Energieübertragung und Hochspannungstechnik
der Universität Stuttgart

2025

Bibliografische Information der Deutschen Nationalbibliothek:

Die Deutsche Nationalbibliothek verzeichnet diese Publikation in der Deutschen Nationalbibliografie, detaillierte bibliografische Daten sind im Internet über http://dnb.dnb.de abrufbar.

Universität Stuttgart
Institut für Energieübertragung und Hochspannungstechnik, Band 48

D 93 (Dissertation Universität Stuttgart)

Objective Interpretation of Frequency Response Analysis of Power Transformers

Publisher: BoD · Books on Demand GmbH, In de Tarpen 42,

22848 Norderstedt, bod@bod.de

Printed by: Libri Plureos GmbH, Friedensallee 273,

22763 Hamburg

ISBN: 978-3-7693-5630-4

I dedicate this thesis to my father,

Rehmat Ullah

Acknowledgements

Glory is to Allah and all Praise be to Him, Almighty, for giving me the ability, health, and knowledge to complete this dissertation and peace be upon Prophet Muhammad (ﷺ) the greatest Lawgiver to the world and a guider for mankind.

I owe my deepest gratitude to my supervisor, Prof. Stefan Tenbohlen, for his continuous support throughout my research. I cannot thank him enough for his guidance, insightful instructions, and encouragement during the past years.

I appreciate Dr. Schärli for the assistance he has given me in administrative matters. My appreciation also goes to all my colleagues in the Grid and High Voltage groups. Those who have left the group and those who are still pursuing their Ph.D's, thank you for your help and for being wonderful friends. I would also like to extend my gratitude to Dr. Satoru Miyazaki for his assistance in the laboratory and for proofreading many of the papers.

I would like to thank all members of Cigré Working Group A2.53 for their valuable support for the collection of fault cases and preparation of a part of the database. I also appreciate the enjoyable discussions within Cigré Working Group meetings.

Not to forget the financial support I received from various sources during my Ph.D. candidature. I thank the Higher Education Commission (HEC) Pakistan, the German Academic Exchange Service (DAAD), and the University of Stuttgart for the financial support during my stay at the Institute of Energy Transmission and High Voltage Technology (IEH).

Finally, I would express a deep sense of gratitude to my parents for their unfailing encouragement and their most generous care, love, and support, which have sustained me always. I wish them a blissful long life. I would also like to express my gratitude to my wife, Zeenat for her support through the ups and downs of this journey and to my son Musab for giving us joy every day. I am also grateful to the rest of my family and friends and my teachers for being supportive and interested in my work during these years.

Abstract

Today the world is changing, and the power grid continues to increase in complexity by integrating intermittent renewables, distributed energy resources, etc. All these factors result in more dynamic loading of the transformer which increases the likelihood of failures. The majority of in-service transformers were installed during the period of rapid economic expansion during the 1980s. Irrespective of the changing generation and demand trends, transformers are expected to last for 30 – 40 years, thus, many transformers are well beyond their intended life. This situation has motivated utilities to implement more efficient maintenance and life expansion strategies in the life cycle of power transformers. These strategies demand proper asset management along with the development of advanced and comprehensive diagnostic methods for transformers.

Among advanced methods, frequency response analysis (FRA) has gained popularity as a sensitive diagnostic test to assess the mechanical integrity of the active part of the transformer. In the last decade, many research efforts have been carried which led to the standardization of FRA measurement procedures. However, the interpretation of FRA results is still backed by the analysis of skillful personnel due to the lack of a standard criteria. Consequently, many international working bodies, i.e., CIGRE, IEEE, and IEC are working in parallel to develop a standard for the condition assessment of the transformer based on FRA results. In contribution to this necessity, this research has proposed some novel methodologies for the objective interpretation of FRA results.

A failure mode analysis is conducted to identify the type of failures in the active part of the transformers. As a consequence, a categorized list of failure modes was compiled. Subsequently, a study on the characterization of the effects of failure modes on FRA was carried out. For this purpose, real case studies are presented and the effects of different failure modes are discussed objectively. Afterward, deviation patterns are generated that summarise the characteristic impact of individual faults on FRA results. Investigation result shows that it is possible to classify different faults based on their deviation patterns in various frequency sub-bands. The failure mode deviation patterns are the basis for the objective interpretation of FRA results. Based on these characterization non-expert users can also gain a better understanding of the difference between two FRA results.

Focusing on the current challenge of developing an effective assessment tool, this study provides a detailed analysis of statistical methods to interpret frequency response measurements. In this research, five criteria for an appropriate numerical index are defined. All indices are evaluated with selected case studies and indices which qualify all criteria are selected as appropriate indices. However, it is not possible to set thresholds for the selected indices as they rely on fixed frequency ranges that can lead to erroneous conclusions. To mitigate these limitations, a vector-based numerical method is defined. Subsequently, a winding assessment factor is introduced and a criterion of abnormality is proposed as an indication of the deformed transformer winding.

This thesis also presents an optimized method to obtain a turn-based 3D high-frequency model of a transformer using the finite integration technique. The novelty of this model is that FRA traces are directly obtained from the 3D model of windings without estimating and solving lumped parameter circuit models. In this model, precise and accurate fault simulations are possible. The model is validated with measurements for the healthy and deformed state of the windings. Good principle agreement of simulation results with the measurements proves the applicability of the model for FRA interpretation studies. The model can be used to generate a large database for different types of winding faults, as it is possible to import a CAD file of the transformer into the software and directly calculate their frequency response without intermediate calculation steps.

In the framework of this research, a simple approach is introduced to estimate the main parameters of the equivalent circuit of a power transformer from different types of FRA measurements. For this purpose, a high-frequency model of a three-phase, three-limb core-type transformer is employed. The effect of transformer equivalent circuit parameters on FRA signature is analyzed through sensitivity studies. To overcome the drawbacks of the conventional indicators, these extracted parameters are then used as identifiers to diagnose fault quantitatively. The performance of the proposed method is analyzed with different case studies.

It is foreseen that the application of intelligent systems will continue to grow in the future and both manufacturers and utilities will start to implement the intelligent systems into their diagnostic instruments. Accordingly, in this doctoral thesis, an expert system for automatic interpretation of FRA has also been developed. In expert system, the numerical indices and machine learning techniques are combined to establish a method for automatic interpretation of FRA. Five machine learning algorithms are studied and real case studies are considered to verify the performance of the expert system. The results suggest that the expert system can reliably identify the six common states of the transformer with good accuracy, and without much human intervention.

In conclusion, the proposed methodologies give insight into the transformer frequency response measurements and ease the objective interpretation of FRA results. Thus, this research provides a way forward towards the establishment of the standard algorithm for a reliable and automatic assessment of transformer FRA results.

Kurzfassung

Mit der Energiewende verändert sich die Welt. Das Stromnetz wird durch die Integration von volatilen erneuerbaren Energien, dezentralen Stromerzeugern immer komplexer. All diese Faktoren führen zu einer dynamischeren Belastung der Leistungstransformatoren, was die Wahrscheinlichkeit von Ausfällen erhöht. Die Mehrheit der in Betrieb befindlichen Transformatoren wurde während der Periode des schnellen wirtschaftlichen Wachstums in den 1980er Jahren installiert. Unabhängig von den sich ändernden Erzeugungs- und Nachfrage-Trends wird erwartet, dass Transformatoren 30 bis 40 Jahre genutzt werden können, sodass viele Transformatoren sich jetzt schon über ihrer geplanten Lebensdauer hinaus befinden. Diese Situation hat die Versorgungsunternehmen dazu motiviert, effizientere Wartungs- und Lebensverlängerungsstrategien im Lebenszyklus von Leistungstransformatoren zu implementieren. Diese Strategien erfordern ein ordnungsgemäßes Asset-Management sowie die Entwicklung fortschrittlicher und umfassender Diagnosemethoden für Transformatoren.

Unter den fortschrittlichen Methoden hat die Frequenzganganalyse (FRA) als empfindlicher Diagnosetest zur Bewertung der mechanischen Integrität des Aktivteils des Transformators an Popularität gewonnen. In den letzten zehn Jahren wurden viele Forschungsanstrengungen unternommen, die zur Standardisierung der FRA-Messverfahren führten. Die Interpretation der FRA-Ergebnisse stützt sich jedoch immer noch auf die Analyse durch Fachexperten aufgrund des Fehlens eines Standardkriteriums. Folglich arbeiten viele internationale Arbeitsgruppen, wie z.B. CIGRE, IEEE und IEC, parallel daran, einen Standard für die Zustandsbewertung von Transformatoren auf Basis der FRA-Ergebnisse zu entwickeln. Als Beitrag zu dieser Notwendigkeit hat diese Forschung einige neuartige Methoden für die objektive Interpretation der FRA-Ergebnisse vorgeschlagen.

Eine Fehlerartenanalyse wird durchgeführt, um die Art der Fehler im Aktivteil des Transformators zu identifizieren. Infolgedessen wurde eine kategorisierte Liste von Fehlerarten erstellt. Anschließend wurde eine Studie zur Charakterisierung der Auswirkungen von Fehlerarten auf die FRA durchgeführt. Zu diesem Zweck werden reale Fallstudien analysiert und die Auswirkungen verschiedener Fehlerarten objektiv diskutiert. Darauf aufbauend werden Abweichungsmuster erstellt, die die charakteristischen Auswirkungen einzelner Fehler auf die FRA-Ergebnisse zusammenfassen. Die Untersuchungsergebnisse zeigen, dass es möglich ist, verschiedene Fehler basierend auf ihren Abweichungsmustern in verschiedenen Frequenzunterbändern zu klassifizieren. Die Abweichungsmuster der Fehlerarten bilden die Grundlage für die objektive Interpretation der FRA-Ergebnisse. Auf Grundlage dieser Charakterisierung können auch unerfahrene Nutzer ein besseres Verständnis für die Unterschiede zwischen zwei FRA-Ergebnissen gewinnen.

Mit Fokus auf die aktuelle Herausforderung der Entwicklung eines effektiven Bewertungswerkzeugs bietet diese Studie eine detaillierte Analyse statistischer Methoden zur Interpretation von Frequenzgangmessungen. Es werden fünf Kriterien für einen geeigneten numerischen Index definiert. Alle Indizes werden mit ausgewählten Fallstudien bewertet, und die Indizes, die alle Kriterien erfüllen, werden als geeignete Indizes ausgewählt. Es ist jedoch nicht möglich, Schwellwerte für die ausgewählten Indizes festzulegen, da sie auf festen Frequenzbereichen basieren, was zu fehlerhaften Schlussfolgerungen führen kann. Um diese Einschränkungen zu mildern, wird ein vektorbasiertes numerisches Verfahren definiert. Anschließend wird ein Wicklungsbewertungsfaktor eingeführt und ein Kriterium der Abnormalität vorgeschlagen, das als Hinweis auf eine deformierte Transformatorwickelung dient.

Diese Dissertation stellt auch eine optimierte Methode zur Erstellung eines 3D-Hochfrequenzmodells eines Transformators unter Verwendung der Finite-Integrations-Technik vor. Die Neuheit dieses Modells besteht darin, dass FRA-Kurven direkt aus dem 3D-Modell der Wicklungen gewonnen werden, ohne geschätzte und gelöste Ersatzschaltkreis-Modelle. In diesem Modell sind präzise und genaue Fehlersimulationen möglich. Das Modell wird durch Messungen für den gesunden und den deformierten Zustand der Wicklungen validiert. Die gute Übereinstimmung der Simulationsergebnisse mit den Messungen beweist die Anwendbarkeit des Modells für FRA-Interpretationsstudien. Das Modell kann verwendet werden, um eine umfangreiche Datenbank für verschiedene Arten von Wicklungsfehlern zu erstellen, da es möglich ist, eine CAD-Datei des Transformators in die Software zu importieren und ohne Zwischenberechnungsschritte direkt deren Frequenzgang zu berechnen.

Im Rahmen dieser Forschung wird ein einfacher Ansatz vorgestellt, um die Hauptparameter des Ersatzschaltbildes eines Leistungstransformators aus verschiedenen Arten von FRA-Messungen zu schätzen. Zu diesem Zweck wird ein Hochfrequenzmodell eines dreiphasigen Transformators mit drei Schenkeln verwendet. Die Auswirkungen der Ersatzschaltbildparameter des Transformators auf die FRA-Signatur werden durch Empfindlichkeitsstudien analysiert. Um die Nachteile der herkömmlichen Indikatoren zu überwinden, werden diese extrahierten Parameter dann als Kennzeichen zur quantitativen Fehlerdiagnose verwendet. Die Leistungsfähigkeit der vorgeschlagenen Methode wird anhand verschiedener Fallstudien analysiert.

Es wird erwartet, dass die Anwendung künstlicher Intelligenz (KI) in der Zukunft weiterwachsen wird und sowohl Hersteller als auch Versorgungsunternehmen beginnen werden, KI in ihre Diagnoseinstrumente zu integrieren. Dementsprechend wurde in dieser Doktorarbeit auch ein Expertensystem zur automatischen Interpretation von FRA entwickelt. Im Expertensystem werden numerische Indizes und maschinelle Lerntechniken kombiniert, um eine Methode zur automatischen Interpretation von FRA zu etablieren. Fünf maschinelle Lernalgorithmen werden untersucht und reale Fallstudien berücksichtigt, um die Leistungsfähigkeit des Expertensystems zu verifizieren. Die Ergebnisse zeigen, dass das Expertensystem zuverlässig die sechs häufigsten Zustände des Transformators mit guter Genauigkeit und ohne großen menschlichen Eingriff identifizieren kann.

Zusammenfassend geben die vorgeschlagenen Methoden Einblick in die Frequenzgangmessungen von Transformatoren und erleichtern die objektive Interpretation der FRA-Ergebnisse. Somit zeigt diese Arbeit einen Weg zur Etablierung eines Standardalgorithmus für eine zuverlässige und automatische Bewertung der FRA-Ergebnisse von Transformatoren.

Contents

List of Figures

List of Abbreviations

Abbreviation	Description
AD	Axial Displacement
AI	Artificial Intelligence
AID	Integral Of Absolute Difference
AI-M	Axial Instability
ANN	Artificial Neural Network
ASLE	Absolute Sum of Logarithmic Error
BUSH-R	Bushing
CART	Classification And Regression Trees
CB-M	Conductor Bending
CC	Correlation Coefficient
CCF	Cross-Correlation Factor
CD	Complex Distance
CDL	Capacitance And Dielectric Loss
CD-M	Core Deformation
CF-M	Clamping Failure
CG	Core Grounding
CGL-E	Core Ground Loss
CIW	Capacitive Inter-Winding
CLS	Concept Learning System
CM	Confusion Matrix
CMD	Condition Monitoring and Diagnostic
CSD	Comparative Standard Deviation
CSI	Cubic Spline Interpolation

CTF-M	Compressive Tension Failure
CT-M	Conductor Tilting
DCWR	DC Winding Resistance
DF	Dissipation Factor
DGA	Dissolved Gas Analysis
DP	Degree Of Polarization
DRA	Dielectric Response Analysis
DSB	Design-Based
DSV	Disk Space Variation
DT	Decision Tree
E	Expectation
ED	Euclidean Distance
EE-OC	End-To-End Open Circuit
EE-SC	End-To-End Short Circuit
EM	Electromagnetic
FEM	Finite Element Method
FIT	Finite Integration Technique
FN	False Negative
FP	False Positive
FRA	Frequency Response Analysis
FRSL	Frequency Response Stray Loss
GI	Gini Index
GUI	Graphical User Interface
HB-M	Hoop Buckling
HF	High Frequency
HFB	High-Frequency Sub-Band

HV	High Voltage
ICA	Independent Components Analysis
ID	Integral Difference
IFDA	Intelligent Fault Detection Algorithm
IF-R	Insulating Fluid
IG	Information Gain
IGR	Information Gain Ratio
IIW	Inductive Inter-Winding
IN	Internal Nodes
JD	Jaccard Distance
LCC	Lin's Concordance Coefficient
LC-M	Loose Clamping
LD-M	Lead Deformation
LFB	Low-Frequency Sub-Band,
LN	Leaf Nodes
LR	Logistic Regression
LSE	Least Squared Error
LV	Low Voltage
MAP	Maximum A Posteriori
MB	Magnetic Balance
MC	Magnetizing Current
MCG-E	Multiple Core Ground
MD	Maximum Of Difference
MFB	Medium Frequency Sub-Band
MI-R	Measurement Instrument
ML	Machine Learning

MM	Minimum Maximum
MSDD	Minimum Value Of SDD
NB	Naive Bayes
NCEPRI	North China Electric Power Research Institute
N-R	Noise
OC-E	Open Circuit Fault
OCOND	Oil Conductivity
PCA	Principal Component Analysis
PC-R	Poor Connection of Short-Circuiting Cables
PD	Partial Discharge
PGP-R	Poor Grounding Practice
POI/DA	Polarization Index and Dielectric Absorption
PRF	Precision, Recall, And F-Score Metrics
RA	Regression Analysis
RD	Radial Deformation
RF	Random Forest
RI-R	Reverse Injection
RM-R	Residual Magnetization
RMSE	Root Mean Square Error
RN	Root Node
RVM	Recovery Voltage Method
SCC-E	Short Circuit Between Conductors
SCCL-E	Short Circuit Between Core Laminations
SCI	Short Circuit Impedance
SCS-E	Short Circuit Between Strands
SD	Standard Deviation

SDA	Standardized Difference Area
SDD	Standard Deviation of Difference
SE	Sum Of Error
SOMO	Saturation Of Moisture in Oil
SSD	Stochastic Spectrum Deviation
SSE	Sum Squared Error
SSMMRE	Sum Squared Min Max Ratio Error
SSRE	Sum Squared Ratio Error
SVD	Singular Value Decomposition
SVM	Support Vector Machine
TA	Thermal Analysis
TCA	Transformer Condition Assessment Algorithm
TDIDT	Top-Down Induction on Decision Trees
TEMP-R	Temperature
TF	Transformer Function
TMB	Time-Based
TN	True Negative
TP	True Positive
TPB	Type-Based
TPD-R	Tap Position Direction
TP-R	Tap Position
TR	Turns Ratio
TS-M	Telescoping
WE-M	Winding Elongation

List of Symbols

Symbol	Description
θ	Arc angle
ρ	Correlation Coefficient
$\sigma(z)$	Sigmoid function
$\varphi(f)$	Phase vector
b	Magnetic flux
b_j	Bias of the hidden layer neuron
b_k	Bias of output layer neuron
C	Number of classes
C_{eq}	Equivalent capacitance
C_g	Ground capacitance
C_{gHv}	Ground capacitance of HV winding
C_{gLv}	Ground capacitance of LV winding
C_{net}	Equivalent network capacitance
C_s	Series capacitance
C_{sHv}	Series capacitance of HV winding
C_{sLv}	Series capacitance of LV winding
C_w	Inter-winding capacitances
D	Data set
d	Dielectric fluxes are allocated on the dual grid.
d	Deformation depth
e	Electric voltages
f	Frequency vector

$f_{parallel-res1}$	First anti-resonance frequency
$f_{parallel-res2}$	Second anti-resonance frequency
f_{res}	Number of data points per decade
F_i	Input feature vector
G	Primary grid
$Gain(D,x)$	Information gain
$\tilde{G}$	Dual grid
$g(x)$	Hyperplane
$H(D)$	Information entropy
h_j	Hidden layer vector
h	Magnetic voltages
L_l	Leakage inductance.
L_e	Exciting inductance
L_m	Magnetizing inductance
MAG	Magnitude of the frequency response
$m(f)$	Magnitude vector
Mu	Mutual inductances between windings
M_u	Mutual inductance
N	Total number of samples
$p(C)$	Prior probability of class.
$p(C\|x)$	Posterior probability of class given predictors
$p(x)$	Prior probability of attributes.
$p(x\|C)$	Probability of predictors given class.
pk	Proportion of samples
r	Radius of the radial deformation
R_e	Core resistance

r_o	Non-deformed radius
R_m	Measurement impedance
R_w	Winding resistance
$TF(f)$	Transfer function
w	Weighting vector
W_{ij}	Weight vector
W_{jk}	Weight vector of hidden and output layer neurons
WS	Window size.
X	Magnitude vector of X TF
$X(i)$	Ith elements of X vector
$\overline{Xw(i)}$	Means of the ith window in X TF
Y	Magnitude vector of Y TF
$Y(i)$	Ith elements of Y vector
y_k	Output vector
$\overline{Yw(i)}$	Mean of the ith window in Y TF
Z_T	Impedance of the winding under test

1 Introduction

Power transformers are one of the most vital assets in electric power transmission and distribution networks and their healthy operation is necessary for the integrity of power systems. The increasing power demand requires these transformers to operate at high load levels. The continuous exposure of power transformers to various system faults give rise to elevated electrical and mechanical stresses that increase the likelihood of failures. The majority of in-service transformers were installed during the period of rapid economic expansion during the 1980s [1]. The planned and expected lifespan of a transformer is generally assumed to be 30-40 years, thus, many transformers are well beyond their intended life and are operated under increasing stress. As load increases, new generation, and economically motivated transmission flows push equipment beyond the nameplate limits. Due to the limited capital investment for new facilities, many transformers are close to or beyond their designed life. As these components age beyond their expected life, there is a risk of an increasing number of catastrophic failures.

There is a focus on the maintenance and life extension of aged transformers to maximize the return on investments. Due to the fact that a lot of substations are approaching to the end of their life, the costs of maintenance and upgrading will increase in the future. Maintenance resulting in lifetime extension is thought to offer important savings by delaying investments [2]. As stated in [3], transformer life extension demands proper asset management along with the development of new and comprehensive diagnostic methods. This suggests that the use of advanced diagnostic methods is a key to extending the life span of transformers.

The precise and timely identification of minor faults is critical. A transformer may continue normal operation under slight damages, however, the capability to withstand further electrical and mechanical stresses is greatly reduced which makes it vulnerable. Any further electrical/mechanical stress may lead to complete failures, power outages, fire, potential injuries, fatalities, and costly repairs. Therefore, early assessment and diagnostic techniques are very important to avoid catastrophic failures of power transformers and ensure the reliable operation of the power network.

During the last decade, many Condition Monitoring and Diagnostic (CMD) techniques have been established to detect faults within power transformers. The dissolved gas analysis (DGA), partial discharge (PD) measurement, thermal analysis (TA), and transformer function (TF) assessment are the main diagnostic methods introduced in the literature [4].

1.1 Motivation

Nowadays, winding deformation is one of the major reasons behind transformer failures. Winding deformations may result from very large electromagnetic forces generated when high short circuit current flows into the windings [5]. Among these winding deformations, radial deformation (Compression and Hoop-tension failure), axial winding elongation “Telescoping”, bulk and localized movement of windings are the frequent mechanical faults. Though the minor winding deformations do not necessarily lead to an immediate transformer failure, however, its mechanical integrity is greatly reduced to withstand future mechanical and electrical stresses. Therefore, the detection of minor winding deformations is very essential to take immediate remedial actions [6].

Transfer function measurement using Frequency Response Analysis (FRA) has drawn attention due to its capability to detect multiple faults in power transformers. Frequency response analysis (FRA) is a sensitive, non-intrusive, and non-destructive tool to determine the mechanical integrity of the active part of the transformer. FRA proved to be useful in diagnosing multiple faults caused during assembly, transportation, installation, and operational life [7], [8], [9], [10]. Due to the inherent significance of FRA as a diagnostic test, many research efforts have been carried out in the last decade which led to the standardization of its measurement procedures as described in IEEE and IEC 60076-18 standards [5], [11]. However, the interpretation of FRA results is still backed by the analysis of skilled personnel due to the lack of standard criteria and automatic algorithms [12]. Thus, the lack of an effective interpretation method is a big challenge to the practical application of FRA in the industry. Consequently, many international working bodies, i.e., CIGRE, IEEE, and IEC are working in parallel to develop a standard for the condition assessment of the transformer based on FRA results [13].

1.2 Current State-of-the-Art

Many recent studies have been focused on the FRA interpretation to detect the extent and the type of mechanical fault. Different algorithms have been proposed for this purpose which can be categorized into three main groups, i.e., simulation models (circuit models/FEM models) [14, p. 1], [15], [16], numerical indices [17], [18], and Artificial Intelligence (AI) techniques [19], [20], [21]. However, these methods have some drawbacks and limitations.

Numerical indices are the most widely used method for transformer frequency response assessment that quantifies the differences in a pair of FRA traces. The final target is a decision about the condition of a transformer based on the quantified values of the indices. Some drawbacks are also associated with numerical indices. Firstly, the results are only numbers, and further interpretations are not possible. Secondly, the definition of the threshold has not been established yet [12], [18]. Moreover, these numerical indices are calculated in fixed frequency sub-bands. Whereas, standards [5], [11] declare that the ranges of frequency sub-bands depend on transformer ratings and cannot be fixed. Due to the lack of effective diagnosis methodology, all the modern FRA equipment also uses fixed frequency sub-bands to compute the numerical indicators for fault detection.

In the literature, circuit models (black box, white box, grey box, etc.) are also employed for interpretation of FRA results [21], [22], [23], [24]. In these circuit models, different sections of transformer windings are represented by circuit elements, such as resistors, inductors, and capacitors, in which the values are extracted either by analytical equations or through finite element methods (FEM). Afterward, the values of these parameters are altered to model and study the impact of various mechanical defects on FRA traces. However, these circuit models have some drawbacks, such as they require an accurate estimation of the winding parameters to model and study the impact of different faults on FRA traces. Secondly, their response is limited to a certain frequency (up to 500 kHz) due to difficulty in solving turn-based parameters. Thirdly, some mechanical changes are difficult to model, and the conversion process involves extra uncertainty. In addition, constant values of the parameters are employed while these parameters are frequency-dependent. Lastly, in all the circuit model studies, the values of one or two parameters are changed to investigate the impact of various mechanical faults, while mechanical deformations like radial deformation, telescoping, and conductor tilting are much more complicated and involve the contribution of more than one parameter.

Artificial intelligence (AI) and Machine Learning (ML) methods have been of great interest lately for diagnosing transformer faults using FRA results and it is gaining popularity due to its effectiveness and high accuracy. Using AI methods, it is possible to classify the changes in the FRA traces and connect them to different fault types and even fault locations [19], [20], [25], [26], [27], [28], [29], [30], [31]. These methods are based on techniques such as genetic algorithms, fuzzy logic, support vector machines, neural network, etc. The major dilemma of using AI methods is the need for a database from real power transformers, such a database usually is not available. Some contributions propose to use the outputs of model experiments and circuit models as a database [19], [32]. However, the applicability of these models for large-scale power transformers is itself an issue.

1.3 Objective

The main objective of this research is the development and optimization of methods for objective and reliable assessment of the mechanical integrity of transformer windings using FRA. These methods can provide an improved understanding of the frequency response measurements and health condition of transformer windings that will facilitate the process of making decisions about their maintenance and replacement.

To achieve this goal, the following intermediate targets were set:

- Data collection of use cases
- Failure mode Analysis
- Development of adaptive frequency division algorithms
- Investigation on the effects of different failure modes in FRA
- Development and evaluation of numerical indicators
- Development of winding assessment factor
- Modelling of transformer frequency response by FEM software
- Transformer equivalent circuit parameter estimation using FRA

- Development of an expert fault detection and classification system based on machine learning classification models

1.4 Thesis Structure

This thesis consists of 8 chapters and 6 appendixes. A brief description of each chapter is given in the following and the relationship between different chapters is provided in Figure 1-1.

Chapter 2

Chapter 2 deals with transformer failure modes and diagnostic techniques with an emphasis on the Frequency Response Analysis (FRA). State-of-the-art approaches for the assessment of FRA measurements are provided along with standard practices and limitations.

Chapter 3

In Chapter 3, the details of the database, adaptive frequency division algorithms, and investigations on the effects of different failure modes on FRA are presented. To summarise the characteristic impact of individual faults on FRA results fault deviation patterns are also provided.

Chapter 4

Chapter 4 deals with the interpretation experience with numerical indices. The evaluation of indices with five criteria is presented in detail. The most appropriate indices are also highlighted. Moreover, a winding assessment factor is proposed for the objective assessment of FRA results.

Chapter 5

Chapter 5 presents 3D High-frequency finite integration modeling and validation with the experimental setup. Simulation of different electrical and mechanical faults is also included in this chapter.

Chapter 6

Chapter 6 presents the transformer equivalent circuit parameter sensitivity studies and methodology for the estimation and extraction of transformer equivalent circuit parameters from FRA measurements.

Chapter 7

Chapter 7 describes an artificial intelligence-based method for automatic condition assessment. The application of six types of machine learning classifiers is briefly discussed. Finally, the development of an expert system based on the selected machine learning model is explained.

Chapter 8

Chapter 8 contains the conclusion and recommendations for future work.

Appendix A

Appendix A contains a list of own publications.

Appendix B

Appendix B presents the application of the adaptive frequency division algorithm for different types and ratings of transformers.

Appendix C

Appendix C summarizes the results of monotonicity of the numerical indices.

Appendix D

Appendix D contains the geometrical data of the windings used for the development of an experimental setup and simulation model.

Appendix E

Appendix E explains the effect of winding structure on FRA measurements. The effect of two common types of windings on FRA is explained in detail.

Appendix F

Appendix F includes the fault simulation results of interwinding connection schemes from high frequency transformer model.

Appendix G

Appendix F presents the theoretical background of machine learning methods and framework of machine learning classifiers.

Appendix H

Appendix F highlights the features and application of the first FRA Graphical User Interface (GUI).

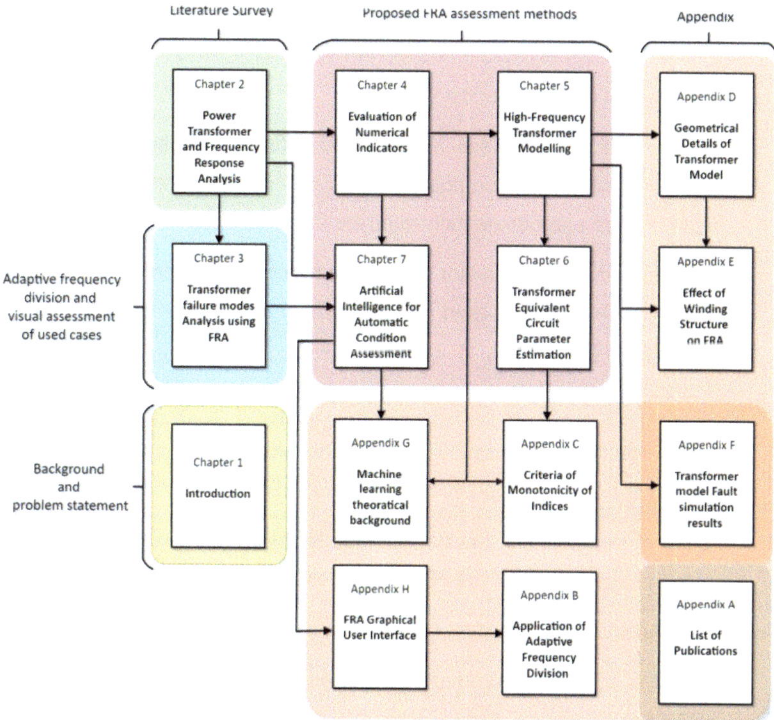

Figure 1-1: Pictorial relationship among chapters and appendixes

1.5 Main Contributions

This section summarizes the main contributions provided by each chapter of this thesis:

Chapter 2: State-of-the-art FRA assessment methods

- Transformer failure modes and fault mechanisms (causes, symptoms, and consequences)
- State-of-the-art FRA interpretation methods and limitations associated with these methods

Chapter 3: Investigation on the effects of failure modes in FRA

- Data collection of use cases in collaboration with CIGRE WG A2.53
- Adaptive frequency division algorithm

- Visual assessment of the effects of different electrical and mechanical faults on FRA

Chapter 4: Evaluation of Numerical Indices

- Several indices are collected on a single platform, compared and categorized with some newly proposed indices based on different criteria, to announce more appropriate indices as the standard ones
- A winding assessment factor (SDD) is introduced and a criterion of abnormality (SDD<-5) is proposed as an indication of the deformed transformer winding

Chapter 5: 3D High-frequency finite integration modeling of transformer FRA

- A 3-phase high frequency (HF) model is developed in contrast to the black-box/white-box circuit models for interpretation of FRA results
- Different electrical and mechanical fault simulations are conducted to discuss the applicability of the HF model in FRA studies

Chapter 6: Transformer Equivalent Circuit Parameter Estimation

- Effects of capacitive parameters are analyzed on resonance frequency in FRA trace
- A mathematical approach is developed to estimate the most relevant parameters of transformer equivalent circuit from FRA measurements

Chapter 7: Application of Artificial Intelligence (AI)

- Performance of six machine-learning classifiers is investigated with real case studies
- An expert system is developed to identify six conditions of the transformer.

Appendix H: Graphical User Interface (GUI)

- A graphical user interface is developed for automatic condition assessment of transformer windings

2 Power Transformer and Frequency Response Analysis (FRA)

In this chapter, a failure mode analysis is presented to categorize possible failure modes in the active part of the transformer. In this analysis, the components that experience issues in the active part of the transformer are identified, together with their causes, mechanisms, and consequences. Subsequently, current state-of-the-art FRA interpretation methods are discussed.

2.1 Power Transformers

A transformer is an electrical system that transfers electrical energy from one circuit to another by electromagnetic induction. Power transformers are a combination of several components such as windings, core, oil-paper insulation system, bushings, on load tap changer, and tank. A careful analysis of the transformer's components is crucial to understand their response under healthy and faulted conditions. The main parts of power transformers are shown in Figure 2-1.

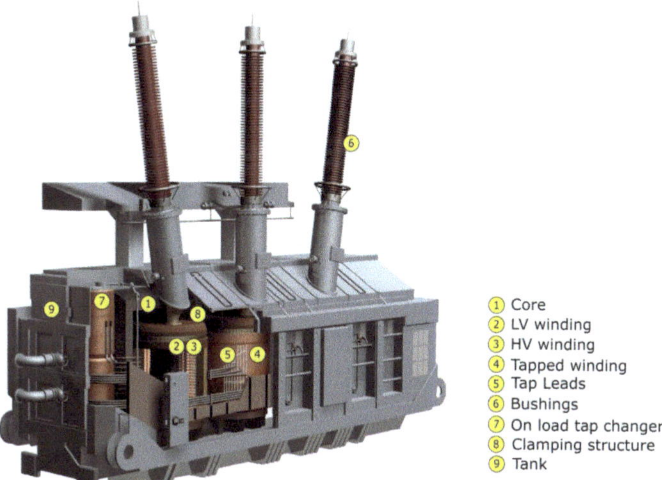

Figure 2-1: Main parts of the power transformer [33]

2.2 Transformer Reliability Surveys

Due to the importance of power transformers in electrical power distribution and transmission systems, several surveys have been conducted for the identification of transformers' reliability in the network. In 2011, CIGRE WG A2.37 collected data from 56 utilities in 21 countries. Data of 964 major failures that occurred from 1996 to 2010 were collected from a population of 23884 transformers of ratings between 69kV and 700kV [34]. The major areas investigated in these surveys are failure mode, failure location, and failure cause.

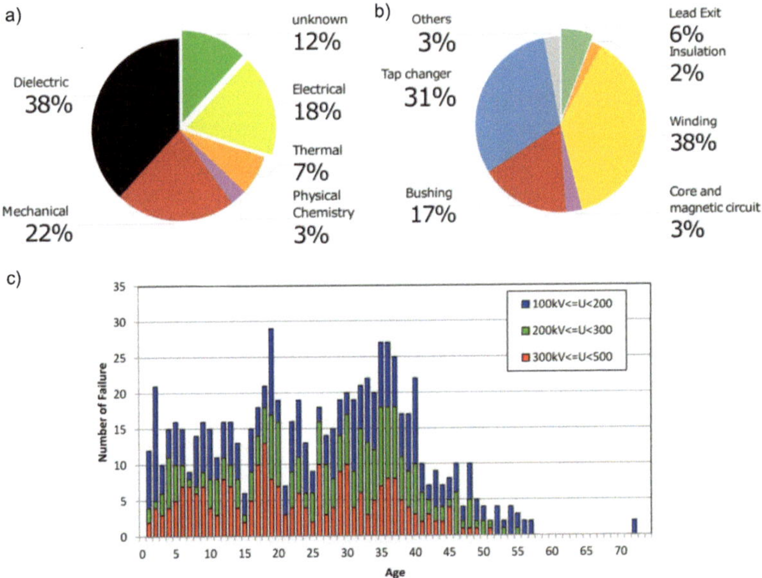

Figure 2-2: Transformer Reliability Survey [34]

a) Failure modes of 799 major failures in substation transformers

b) Failure location of 536 major failures in substation transformers

c) Number of transformer failures with age

Figure 2-2 illustrates the key findings of the transformer reliability survey. Based on Figure 2-2, insulation damage/dielectric, mechanical faults, and electrical faults are the main reasons for transformer failure. It can also be seen in Figure 2-2 that the transformer failure location, in most cases is at windings where 38% of faults occur, followed by 31% at tap changer and 17% at bushings.

According to the Transformer Reliability Survey Cigre A2.37, the failure rate depends on transformer age as shown in Figure 2-2. Thus, monitoring and maintenance becomes more important as transformer ages. Worldwide electric power demand rose sharply during the period of rapid economic expansion (1970-1980), and oil-immersed transformers and other electric power equipment were manufactured in very large quantities during this time. At present, most of the installed transformers have passed their expected lifespan of 30 years. Hence, the number of aging transformers is expected to increase in the future.

2.3 Transformer conditions during the life cycle

During the life cycle, a transformer undergoes different conditions, i.e., normal, defective, faulty, and failed, as suggested by [1]. Transformer typical aging curve with corresponding conditions is shown in Figure 2-3. The definition of different stages is presented in Table 2-1. From the aging curve, it can be seen that the withstand strengths of a transformer naturally decrease with time due to various aging processes and operational stresses (normal aging) but may deteriorate faster than normal under the occurrence of an external fault (accelerated aging).

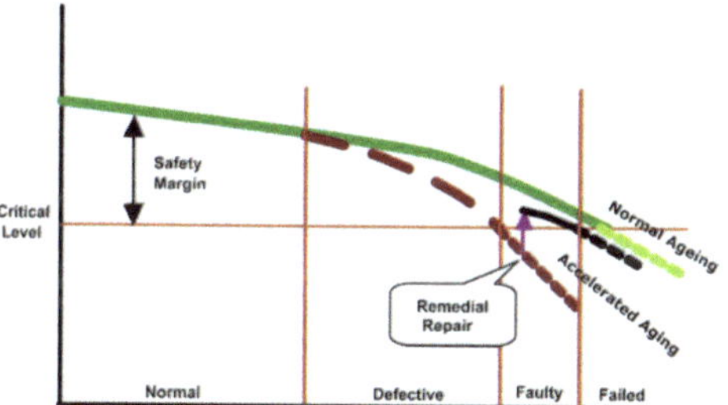

Figure 2-3: Transformer aging curve [1]

Table 2-1: Conditions of a transformer in the course of its life cycle [1]

Transformer Condition	Description
Normal	No obvious problems. No evidence of degradation
Defective	No significant impact on short-term reliability, but asset life is adversely affected in the long term unless remedial action is taken
Faulty	Can remain in service, but short-term reliability likely to be reduced
Failed	Cannot remain in service. Remedial action is required before equipment can be returned to service

2.4 Transformer Winding Failure Modes

Referring to the transformer reliability survey [34], windings are involved in a large portion of transformer failures. It is also assumed that some dielectric failures are initiated by mechanical movements inside the winding, and these could have been avoided by assessing the mechanical condition of the winding and core at an early stage. The mechanical integrity of a transformer winding is challenged by several mechanisms, i.e., mechanical forces due to excessive short-circuit currents, excessive mechanical acceleration during transport (accidents), dynamic forces in service (vibration or seismic forces), aging which reduce clamping force on supportive structure, etc. These mechanisms initiate different failure modes that give rise to the different types of faults in transformer windings. Transformer winding failure modes can be classified into two groups, mechanical, and electrical failure modes. These are briefly explained in the following sections.

2.4.1 Mechanical Failure Mode

Transformer winding mechanical faults are typically caused due to mechanical and/or electrical changes inside the transformer. Mechanical changes can arise during short-circuit events (generating large electromagnetic forces), transportation, installation and/or service life, etc. Among these, the failures due to the generation of electromagnetic forces are the most common. The electromagnetic forces are resolved into two components, i.e., radial and axial. The radial (axial) forces are generated by the interaction of the short-circuit current and axial (radial) components of the leakage flux density, as shown in Figure 2-4.

These massive electromagnetic forces can displace or deform the windings. The radial forces act inwards on the inner winding producing compression failure, and they act outwards on the outer winding producing hoop tension failure [35]. While the axial forces are generated parallel to the height of the winding, and they act in the opposite direction due to the pattern of the magnetic field. These opposing forces can be of two types, i.e., axial compression forces and axial expansion forces. Axial compression forces are directed axially toward the winding centers and can collapse the winding. Axial expansion forces are directed axially toward the top and bottom clamping plates and can bend or break them, allowing axial instability [35]. While electrical changes, i.e., short-turn, open circuit, can be caused due to aging and/or subsequent mechanical failure. The detailed mechanism of mechanical failure modes along with the causes and consequences is described in Figure 2-5.

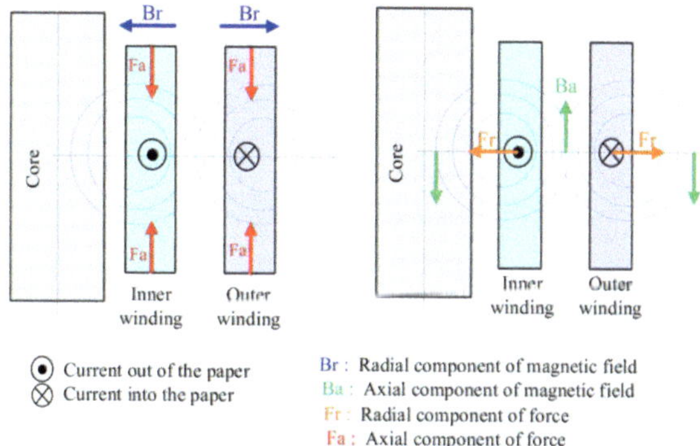

Figure 2-4: Axial (left) and radial (right) electromagnetic forces generated by the interaction of windings current and leakage flux

The most typical mechanical failures in transformers are summarized in Table 2-2. Among these, seven failures (conductor tilting, winding elongation, conductor bending, telescoping, axial instability, loose clamping, and clamping failure) belong to the family of axial deformations. Figure 2-6 shows the example of the failure modes in transformer windings due to axial forces. Two kinds of failures (hoop buckling and compression failure) belong to the family of radial deformations. Figure 2-7 shows the example of failure modes in transformer winding due to radial forces. Figure 2-8 shows the core damage fault during transportation accidents.

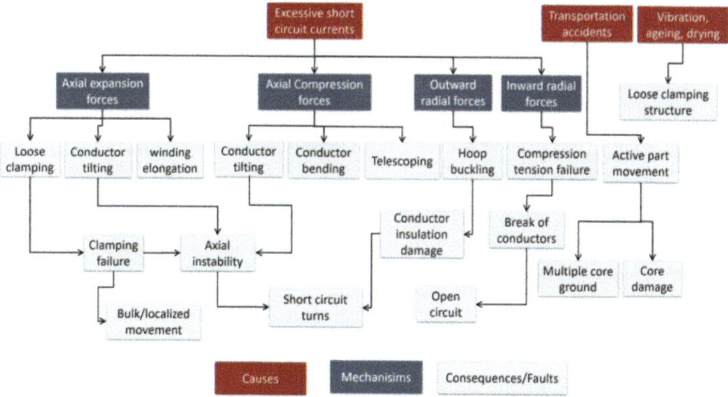

Figure 2-5: Causes and mechanisms leading to the mechanical failure in transformer windings

Table 2-2: Common mechanical failures in power transformers

Abbreviation	Failure mode
CT-M	Conductor tilting
WE-M	Winding elongation,
CB-M	Conductor bending
TS-M	Telescoping
AI-M	Axial instability
LC-M	Loose clamping
CF-M	Clamping failure
HB-M	Hoop buckling
CTF-M	Compressive tension failure
CD-M	Core deformation
LD-M	Lead deformation

Figure 2-6: Examples of the failure modes in transformer windings due to axial forces [13]
(a) Conductor tilting (b) Conductor bending (c) Axial instability (d) Loose clamping
(e) Winding elongation (f) Telescoping

a)

b)

Figure 2-7: Examples of the failure modes in transformer windings due to radial forces [13]
(a) Compressive tension failure
(b) Hoop buckling

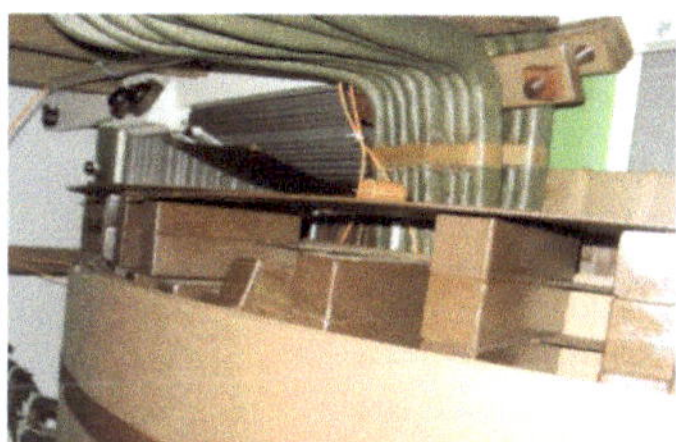

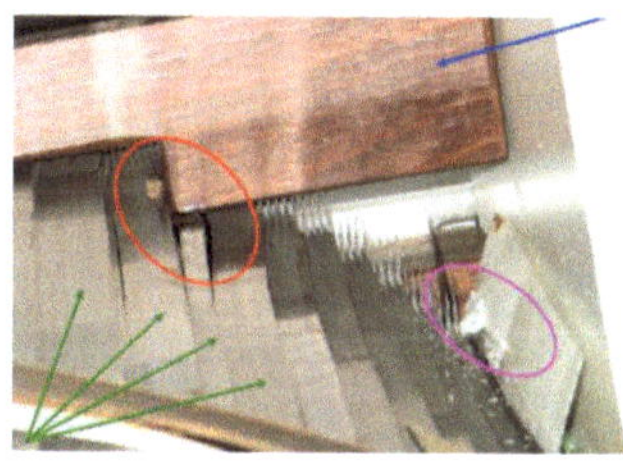

Figure 2-8: Examples of core damage fault during transportation accidents [13]

2.4.2 Electrical Failure Modes

The most common electrical failure modes in transformer winding and core are classified into six types, i.e., short circuit between strands (SCS-E), short circuit between turns (SCT-E), open circuit (OC-E), core ground loss (CGL-E), multiple core ground (MCG-E), and short circuit between core laminations (SCCL-E) as listed in Table 2-3. The detailed mechanism of electrical failure modes along with the causes and consequences is described in Figure 2-9. Under the influence of SCS-E, SCT-E, OC-E, and SCCL-E faults a transformer cannot remain in service and is considered to be in the stage "Failed" of its life cycle. While during CGL-E and MCG-E faults a transformer can remain in service but it is considered to be in the “Faulty” stage. However, the overheating created by circulating currents during MCG-E will lead to a degradation of the insulation between core laminations and could result in the SCCL-E fault. Figure 2-10 shows some examples of the electrical failure modes in transformer windings and core.

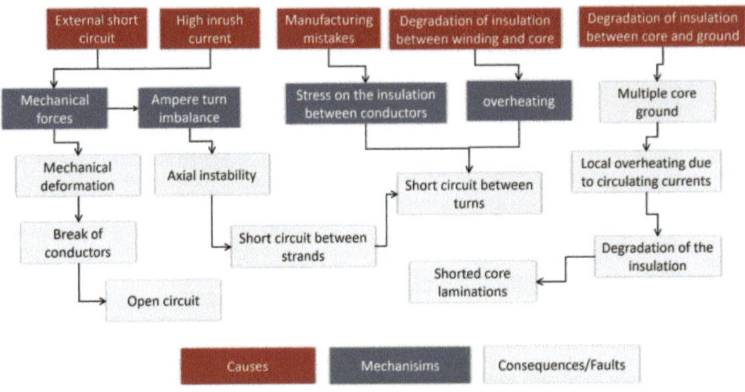

Figure 2-9: Causes and mechanisms leading to the electrical failures in transformer windings

Table 2-3: Common electrical failure modes in power transformers

Abbreviation	Failure mode
SCS-E	Short circuit between strands
SCC-E	Short circuit between conductors
OC-E	Open circuit
SCCL-E	Short circuit between core laminations
CGL-E	Core ground loss
MCG-E	Multiple core ground

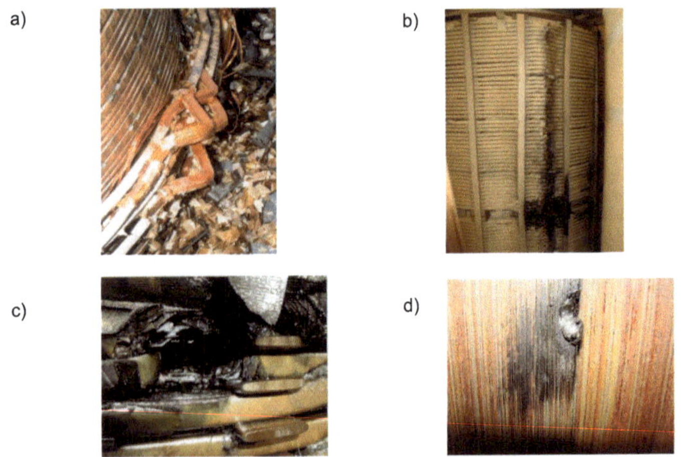

Figure 2-10: Examples of the electrical failure modes in transformer windings and core [13]
(a) Bending and short circuit of conductors (b) Buckling and short circuit of conductors (c) Open circuit of lead conductor (d) shorted core laminations

2.5 Transformer Condition Monitoring and Diagnostic Methods

Different condition monitoring and diagnostic methods have been developed to recognize different faults in the active part of the transformer. These methods can be classified as conventional and as advanced as appreciated in Figure 2-11. While conventional methods can be divided into two categories based on the type of the method, i.e., electrical and chemical. Among the electrical methods, the most popular methods are transformer turn ratio (TR), magnetizing current (MC), magnetic balance test (MB), DC winding resistance (DCWR), dissipation factor (DF), insulation resistance (IR), polarization index and dielectric absorption (POI/DA) and short-circuit impedance (SCI). If the core grounding strap is available, the insulation resistance of the core to the ground can also be measured (CG). In the category of chemical methods, the following most popular methods are dissolved gas analysis (DGA), furan analysis (FA), degree of polymerization (DP) of paper, measurements of moisture content in paper (SOMO), and oil conductivity measurements.

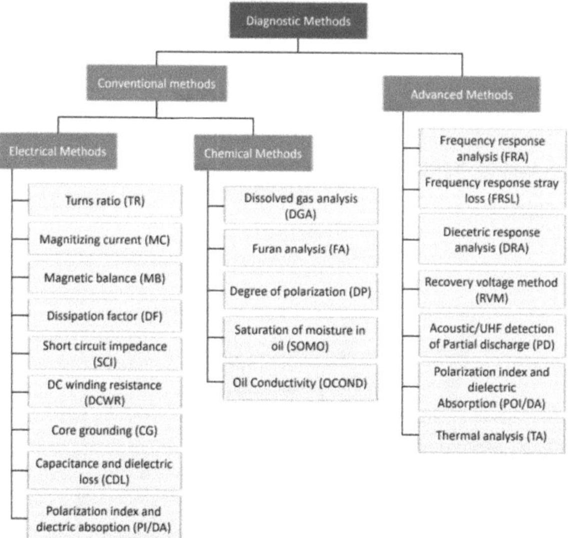

Figure 2-11: State-of-the-art transformer condition monitoring and diagnostic methods

Among advanced methods, frequency response analysis (FRA) has received great attention due to its ability to detect a variety of mechanical deformations.

The frequency response of stray losses (FRSL) has also been used in some utilities for the diagnosis of short-circuits between parallel strands and local overheating due to excessive eddy current losses. Other advanced methods are partial discharge measurement (PD) and thermal analysis (TA).

2.6 Frequency Response Analysis

Frequency response analysis is a widely accepted, non-intrusive, and sensitive method to determine the mechanical integrity of the active part (windings, leads, and core) of the transformer. It was first tested on field transformers by Dick and Erven in 1978 [36]. It is a non-destructive, offline test, which measures the transfer function (TF) of the transformer winding from a few Hz to several MHz. FRA is a comparative diagnostic method that requires a reference measurement also known as a 'fingerprint'. Time-based, type-based, and design-based comparisons can be performed to assess the mechanical condition of the transformer [5], [11].

In FRA, transfer functions (TFs) of transformers are measured. TFs are determined by the electrical parameters of power transformers, such as capacitance and inductance of windings, capacitance between windings, capacitance between windings and ground, etc. This diagnostic method is based on the principle that the geometrical changes in the transformer core and windings due to mechanical stresses can be considered as a change in the RLC parameters of the equivalent circuit of the power transformer. Therefore, these changes can be detected through a change in the TF of the transformer.

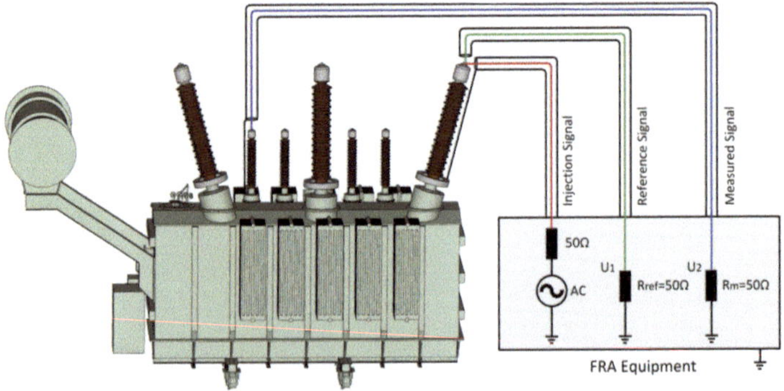

Figure 2-12: Frequency response analysis (FRA) measurement setup

The behavior of the measured TF can reveal different electrical and mechanical changes in the physical geometry of the transformer. An FRA test is conducted offline on a de-energized transformer. The measurement is performed over a wide range of frequencies (a few Hz to several MHz), and the results are compared with a reference measurement "fingerprint" of the winding to perform a diagnosis. The measurement setup for the FRA method is shown in Figure 2-12. A voltage signal of variable frequency is injected at one terminal, and the response is measured at the other terminal. The response will vary in magnitude and phase.

From the measured voltages $U1$ and $U2$, the transfer function $TF(f)$ is computed according to equation (2-1). Through circuit analysis, it can be demonstrated that the $TF(f)$ is proportional to R_m (which is usually 50 Ω) and inversely proportional to R_m plus the impedance of the winding under test *(Z_T)*. The results of TF measurements can be represented in several ways. In the field, FRA measurements are represented in the "bode plot" representation. This representation consists of two graphs: magnitude *(m)* plot and phase *(φ)* plot, where m and φ are calculated by Equation (2-2) and (2-3).

$$TF(f) = \frac{U_2(f)}{U_1(f)} = \frac{R_m}{R_m + Z_T(f)} = \frac{50}{50 + Z_T(f)} \qquad (2\text{-}1)$$

$$m(f) = 20\log\frac{U_2(f)}{U_1(f)} \qquad (2\text{-}2)$$

$$\varphi(f) = \tan^{-1}(\angle U_2(f) - \angle U_1(f)) \qquad (2\text{-}3)$$

2.6.1 FRA Test Configurations

The frequency response of a transformer has a fundamental relationship with different physical components of the transformer, i.e., core, interaction between windings, winding structure, etc. Consequently, different frequency response characteristics can be expected for different test configurations. In the existing recommended practices [5] and [11], mainly four types of test configurations have been standardized. These configurations are End-to-end open circuit (EE-OC), End-to-end short circuit (EE-SC), capacitive inter-winding (CIW), and inductive inter-winding (IIW) as appreciated in Figure 2-13.

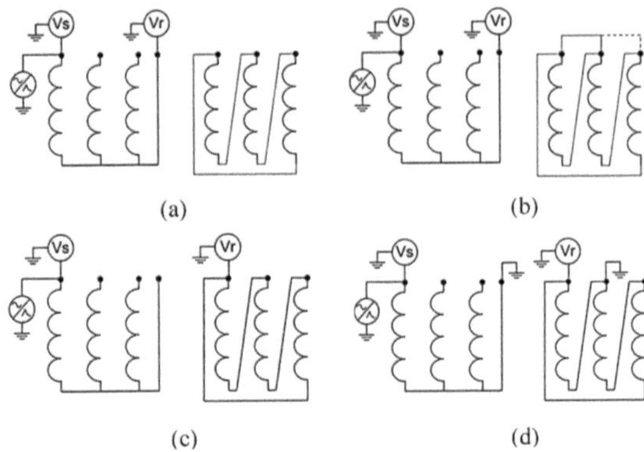

Figure 2-13: Connection schemes for FRA measurements
(a) end-to-end open circuit (EE-OC)
(b) end-to-end short circuit (EE-SC)
(c) capacitive inter-winding (CIW)
(d) inductive inter-winding (IIW)

2.6.2 FRA Test Comparison Practices

As FRA is a comparative diagnostic method, consequently, three types of comparison are practiced in the field, i.e., time-based (TMB), type-based (TPB), and design-based (DSB) comparison as illustrated in Figure 2-14. In time-based comparison, the current FRA trace is compared with the previously measured FRA trace from the same transformer unit.

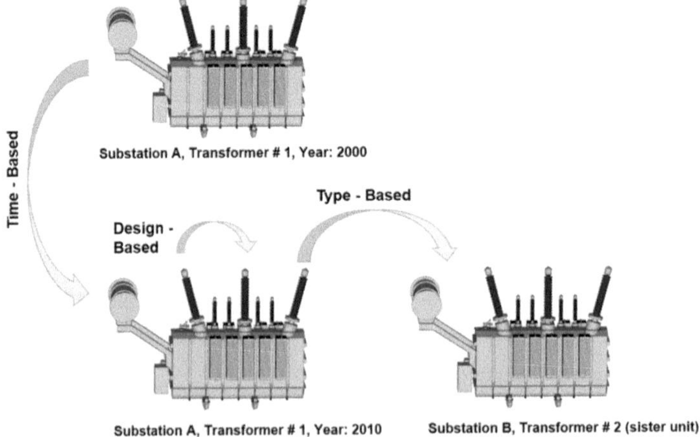

Figure 2-14: FRA test comparison practices

In type-based comparison, the FRA trace of one phase is compared with the trace of other phases of the same transformer unit. In the case of design-based comparison, FRA traces of two sister units are compared. As a general practice, experts first try to carry out a TMB comparison and in case they find suspicious deviations then TPB comparison is conducted as a complement to TMB comparisons. However, if TPB is also not possible especially for single-phase transformer units or for those units that possess asymmetries then experts look for sister unit comparisons (DSB). These are general practices that define the acting conduct of experts in the field of FRA. A typical time-based comparison of EE-OC FRA traces is shown in Figure 2-15.

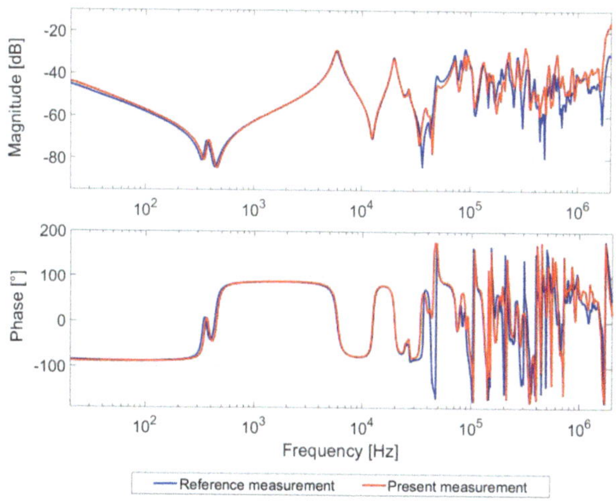

Figure 2-15: Example of a time-based comparison of EE-OC FRA before and after fault

2.7 Current State-Of-The-Art FRA Interpretation Methods

The current state-of-the-art FRA interpretation methods are summarized in Figure 2-16. As can be appreciated the existing methods can be classified into three groups: quantitative numerical indicators, simulation models, and intelligent systems based on machine learning and pattern recognition algorithms. Following, a summary of the studies found in the literature about these methods is presented and discussed.

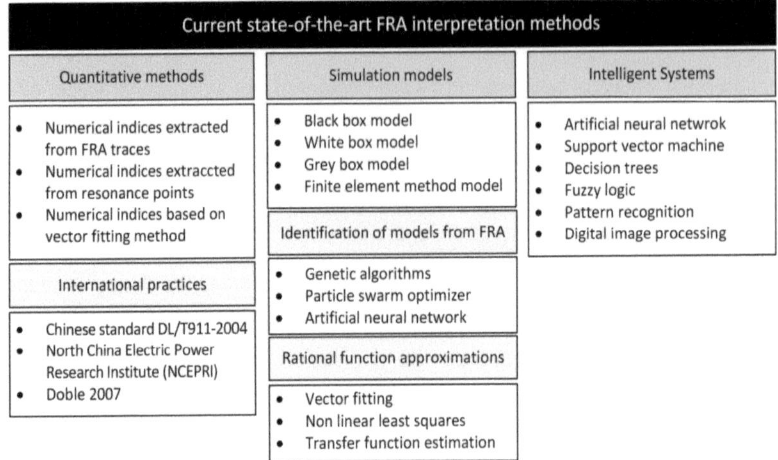

Figure 2-16: Summary of the state-of-the-art FRA interpretation methods

2.7.1 Quantitative Indicators

The interpretation of FRA results using numerical indices is based on the fact that the numerical indices quantify the deviation between reference and present TFs. In this method, a mathematical equation is used to extract a single value from the reference and present TFs. The assessment of the transformer condition is then carried out based on the quantified value of the indices. The values that are extracted using these equations or indices are precise and repeatable [18]. It is noteworthy that the previous contributions use different frequency ranges for interpretation purposes. Table 2-4 summarizes the frequency ranges used in the literature for the FRA interpretation. However, ref [5], [11] declare that frequency sub-bands are not standardized, and a general range cannot be concluded.

In literature the application of numerical indices is carried out in three different methods, i.e., indices calculated directly from FRA traces, indices calculated from resonance and anti-resonance points, and indices calculated from rational functions. In this research, only the indices calculated directly from the traces are presented. The definitions of these indices are described in Table 2-5.

Table 2-4: Frequency bands used in literature for FRA interpretation

Ref.	Frequency range	Ref.	Frequency range	Ref.	Frequency range
			Single frequency sub-band		
[37], [38]	20 Hz–2 MHz	[19], [39]	20 Hz-1 MHz	[40]	10 Hz–1 MHz
[10], [41]	100 kHz–1 MHz	[42], [43]	10 Hz–3 MHz	[44]	10 kHz–4 MHz
[4], [7]	5 kHz–1 MHz	[4], [7]	5 kHz–2 MHz	[32]	10 kHz–1 MHz
			Three frequency sub-bands		
[17], [45]	0–100 kHz 100–600 kHz 600–1000 kHz	[46]	0–20 kHz 20–400 kHz 400–1 MHz	[47]	300 Hz–50 kHz 50 kHz–1 MHz 1–3 MHz
[48]	20 Hz–10 kHz 10–100 kHz 100 kHz–1 MHz	[49]	100 Hz–20 kHz 20–200 kHz 200 kHz–2 MHz	[50]	0–350 kHz 350 kHz–1 MHz 1–2 MHz
			More than three frequency sub-bands		
[51]	100 Hz–1 kHz 1–10 kHz 10–100 kHz 100 kHz–1 MHz	[5]	0–2 kHz 2–20 kHz 20 kHz–1 MHz 1–2 MHz	[52]	10 kHz–1 MHz 1–2 MHz 2–3 MHz 3–5 MHz
[53]	0–2 kHz 2–20 kHz 20–400 kHz 400 kHz–1 MHz	[54]	1–10; 10–20 kHz 20–40; 40-100 kHz 100–500 kHz 500–1000 kHz	[55]	20 Hz–1 MHz 10 regions

2.7.1.1 Numerical indices calculated directly from FRA traces

In this method mostly the entire FRA spectrum is divided into different sub-bands and the assessment of the transformer is then carried out based on the final value from each sub-band as depicted schematically in Figure 2-17.

In the literature, numerous indices have been proposed as described in Table 2-5. The equations of these indices are summarized here. In the following equations, X and Y are the magnitude vectors of the reference and present TFs, respectively, $X(i)$ and $Y(i)$ are the *ith* elements of these vectors, *f* is the frequency vector, *N* is the total number of samples, and Φ_X and Φ_Y are the phase vectors of the TFs. The values of the amplitude vectors used for calculating these indices can be in their original form or the dB scale. In the literature, a dB scale is usually used. However, one can also use the original values, which leads to different results than when using dB scale values. It is noteworthy that there is no report in the literature about the advantages of each approach.

Table 2-5: Abbreviation definition of the numerical indices along with their reference

Abbr.	Equations	Definition	Reference
ED	Equation (2-4)	Euclidean Distance	[32], [55]
CD	Equation (2-5)	Complex Distance	[56]
SD	Equation (2-6)	Standard Deviation	[42]
ID	Equation (2-7)	Integral Difference	[43], [57]
AID	Equation (2-8)	Integral of Absolute difference	[32], [57]
SDA	Equation (2-9)	Standardized Difference Area	[17]
ASLE	Equation (2-10)	Absolute Sum of Logarithmic Error	[42], [47]
RMSE	Equation (2-11)	Root Mean Square Error	[39]
E	Equation (2-12)	Expectation	[38]
SSD	Equation (2-13)	Stochastic Spectrum Deviation	[58]
MD	Equation (2-14)	Maximum of Difference	[43], [55]
CCF	Equation (2-15)	Cross-Correlation Factor	[17], [38]
CC	Equation (2-16)	Correlation Coefficient	[37], [42]
SSE	Equation (2-17)	Sum Squared Error	[47], [55]
SSRE	Equation (2-18)	Sum Squared Ratio Error	[32], [55]
SSMMRE	Equation (2-19)	Sum Squared Min Max Ratio Error	[47]
CSD	Equation (2-20)	Comparative Standard Deviation	[59]
LCC	Equation (2-21)	Lin's Concordance Coefficient	[59]
SE	Equation (2-22)	Sum of Error	[59]
LSE	Equation (2-23)	Least Squared Error	[51]
MM	Equation (2-24)	Minimum Maximum	[54]
JD	Equation (2-25)	Jaccard Distance	[60]

Reference [55] introduces *ED* and compares it with numbers of other indices. It emphasizes that the linearity of the indices versus the severity of the fault is important. Reference [56] also calculates *ED* in different frequency ranges and implements them for defining the fault type. Reference [42] uses *CC*, *ASLE,* and *SD* for detecting the axial and radial deformation faults in three large transformers. It reports that all three indices can show a fault though *CC* fails to detect axial displacements less than 1% of the winding height. Therefore, it is a less sensitive index. This reference employs the indices for the phase and sister unit comparison and explains that fault detection is possible only for severe faults. It also takes 1% of axial displacement to set criterion limits for the indices.

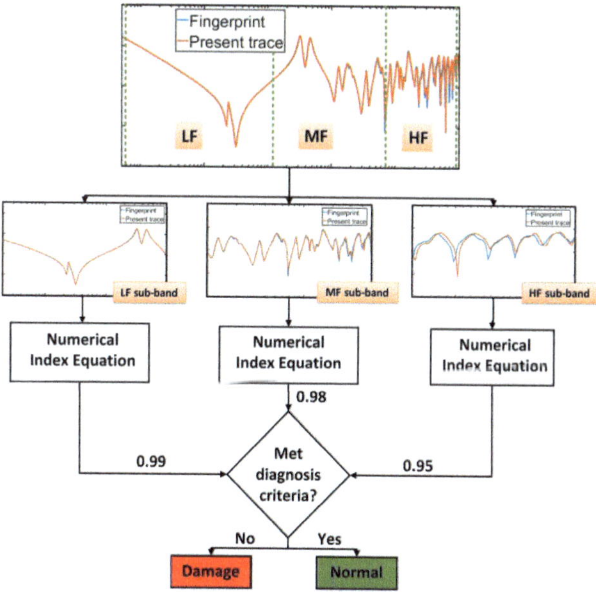

Figure 2-17: Application of a numerical index for FRA interpretation

$$ED = \sqrt{(X-Y)^T(X-X)} = \sqrt{\sum_{i=1}^{N}(Y(i)-X(i))^2} \tag{2-4}$$

$$CD = \sqrt{\sum_{i=1}^{N} \begin{matrix} [(X(i)cos\varphi_X(i) - Y(i)cos\varphi_Y(i))^2 \\ +(X(i)cos\varphi_X(i) - Y(i)cos\varphi_Y(i))^2] \end{matrix}} \tag{2-5}$$

$$SD = \sqrt{\frac{\sum_{i=1}^{N}(Y(i)-X(i))^2}{N-1}} \tag{2-6}$$

$$ID = \int ((Y(f) - X(f))\ df \tag{2-7}$$

$$AID = \int |((Y(f) - X(f))|\ df \tag{2-8}$$

$$SDA = \frac{\int |((Y(f) - X(f))|\ df}{\int |X(f))|\ df} \tag{2-9}$$

$$ASLE = \frac{\sum_{i=1}^{N}|20\log_{10}Y(i) - 20\log_{10}X(i)|}{N} \tag{2-10}$$

$$RMSE = \sqrt{\frac{1}{N}\sum_{i=1}^{N}\left[\frac{|Y(i)| - |X(i)|}{\frac{1}{N}\sum_{i=1}^{N}|X(i)|}\right]^2} \tag{2-11}$$

$$\Delta(i)=\frac{|Y(i)|-|X(i)|}{\frac{1}{N}\sum_{i=1}^{N}|X(i)|}$$

$$E[\Delta]=\frac{1}{N}\sum_{i=1}^{N}\Delta(i) \tag{2-12}$$

$$SSD=\frac{100}{N}\sum_{i=1}^{N}\frac{|Y(i)-X(i)|}{X(i)} \tag{2-13}$$

$$MD=\max(Y(i)-X(i)) \tag{2-14}$$

$$CCF=\frac{\sum_{i=1}^{N}(X(i)-\bar{X})(Y(i)-\bar{Y})}{\sqrt{\sum_{i=1}^{N}(X(i)-\bar{X})^2\sum_{i=1}^{N}(Y(i)-\bar{Y})^2}} \tag{2-15}$$

$$CC=\frac{\sum_{i=1}^{N}X(i)Y(i)}{\sqrt{\sum_{i=1}^{N}\left(X(i)\right)^2\sum_{i=1}^{N}\left(Y(i)\right)^2}} \tag{2-16}$$

$$SSE=\frac{\sum_{i=1}^{N}(Y(i)-X(i))^2}{N} \tag{2-17}$$

$$SSRE=\frac{\sum_{i=1}^{N}\left[\frac{Y(i)}{X(i)}-1\right]^2}{N} \tag{2-18}$$

$$SSMMRE=\frac{\sum_{i=1}^{N}\left[\frac{\max\,(X(i),Y(i))}{\min\,(X(i),Y(i))}-1\right]^2}{N} \tag{2-19}$$

$$CSD=\frac{\sum_{i=1}^{N}[(Y(i)-\bar{Y})-(X(i)-\bar{X})]^2}{N-1} \tag{2-20}$$

$$LCC=\frac{2Sxy}{(Y-X)^2+Sy^2+Sx^2}$$

$$Sx^2=\frac{1}{N}\sum_{i=1}^{N}(X(i)-\bar{X})^2,\ Sy^2=\frac{1}{N}\sum_{i=1}^{N}(Y(i)-\bar{Y})^2 \tag{2-21}$$

$$Sxy=\frac{1}{N}\sum_{i=1}^{N}(X(i)-\bar{X})(Y(i)-\bar{Y})$$

$$SE=\frac{\sum_{i=1}^{N}(Y(i)-X(i))}{N} \tag{2-22}$$

$$LSE=\frac{\sum_{i=1}^{N}(X(i)-\bar{X})(Y(i)-\bar{Y})}{\sum_{i=1}^{N}(X(i)-\bar{X})^2} \tag{2-23}$$

$$MM=\frac{\sum_{i=1}^{N}min\left(Y(i),X(i)\right)}{\sum_{i=1}^{N}max\left(Y(i),X(i)\right)} \tag{2-24}$$

$$JD=\frac{\sum_{i=1}^{N}(X(i)-Y(i))^2}{\sqrt{\sum_{i=1}^{N}\left(X(i)\right)^2}-\sqrt{\sum_{i=1}^{N}\left(X(i)\right)^2-\sum_{i=1}^{N}X(i)Y(i)}} \tag{2-25}$$

2.7.1.2 International Practices

In literature, some efforts are also made to set limits/thresholds to these numerical indices such as Chinese standard DL/T911-2004 [45], North China Electric Power Research Institute (NCEPRI) difference index [61], and Doble CCF criteria [46].

2.7.1.2.1 Chinese Standard DL/T911-2004

The Chinese standard has defined a criterion to detect deformation inside the transformer based on the correlation coefficient (*CC*) [45]. In this standard, three frequency ranges have been described. The interpretation is based on a relative factor, R_{xy}, which is defined using Equation (2-1).

$$Rxy = \begin{cases} 10 & 1-\rho < 10^{-10} \\ -\log_{10}(1-\rho) & others \end{cases} \qquad (2\text{-}1)$$

Where ρ is calculated using *(2-16)*.The frequency ranges and limits for the relative factor are given in Table 2-6.

Table 2-6: Criterion for FRA assessment based on R_{xy} [45]

Category	Limits for the Relative factor (R_{xy})
Severe	($R_{LF} < 0.6$)
Obvious	($1.0 > R_{LF} \geq 0.6$) or ($R_{MF} < 0.6$)
Slight	($2.0 > R_{LF} \geq 1.0$) or ($0.6 \leq R_{MF} < 1.0$)
Normal	($R_{LF} \geq 2.0$), ($R_{MF} \geq 1.0$) and ($R_{HF} \geq 0.6$)
Frequency Ranges: LF: 1-100 kHz, MF: 100-600 kHz, HF: 600-1000 kHz	

2.7.1.2.2 North China Electric Power Research Institute (NCEPRI)

The NCEPRI published an algorithm based on a difference index "E" to assess the condition of windings [61]. The frequency range in which this index is calculated depends on the rating of the winding under test. The difference index is defined using Equation (2-2). The threshold values of E are shown in Table 2-7.

$$E = \frac{1}{N}\sum_{i=1}^{N}(X(i) - Y(i))^2 \qquad (2\text{-}2)$$

Table 2-7: Criterion for FRA assessment based on E [61]

Category	Limits for the factor E
Normal condition	E < 3.5
Slight distortion	3.5 < E < 7.0
Severe distortion	E > 7.0
Frequency Ranges: HV: 10-515 kHz, LV: 10-600 kHz, TV: 10-700 kHz	

2.7.1.2.3 Doble CCF criteria

In 2007, Doble provided a Cross-Correlation Factor (CCF) and developed the interpretation criterion for time-based comparison [46]. CCF is calculated using Equation (2-15). Threshold values of CCF are shown in Table 2-8.

Table 2-8: Criterion for FRA assessment based on CCF [46]

Category	Limits for the CCF
Investigate	< 0.95
Marginal	0.96-0.97
Good	0.98-1.00
Frequency Ranges: LF1: 20Hz-2kHz, LF2: 2-20kHz, MF: 20-400kHz, HF: 400-1000kHz	

2.7.1.2.4 Summary on International Practices

Some drawbacks associated with the proposed international criteria are listed below:

- Existing interpretation criteria are based on fixed frequency sub-bands. While frequency sub-bands cannot be generalized this makes the existing criteria unreliable in many cases.
- Existing interpretation criteria do not speak about the type of fault during the FRA assessment
- Existing interpretation criteria were extracted from a small population of transformers.
- Existing interpretation criteria are only valid for time-based comparisons. There is a lack of an approach for other comparison methods.

Due to these drawbacks, many cases reported unsuccessful performance of these criteria. Ref [37], [38], [59] report cases where the transformers have obvious deformations, but the Chinese standard fails to show deformations, indicating a "normal winding". Similarly, ref. [59], [62] reported two cases where CCF could not detect any fault. Therefore, such inconsistencies show that these criteria need revision for better interpretation of FRA results. As the application of general fixed frequency sub-bands is only a gross estimation that is valid in many cases, but not in other cases. Due to the lack of an effective diagnosis method, all the modern FRA equipment also uses fixed frequency sub-bands to compute the indicators for fault detection.

2.7.2 Simulation Models

In literature, the FRA interpretation problem is also addressed by developing different physical models. Accordingly many, circuit models have been developed for the interpretation of transformer frequency responses [21], [22], [23], [24]. The basic idea behind the physical models is to replicate the behavior of transformer windings. However, the modeling of transformer winding is a compromise between accuracy and complexity. Transformer winding models for FRA are categorized into three groups: white box, black box, and grey box models. Each category is suitable for certain situations and possesses its own advantages.

2.7.2.1 White box model

In white-box models, different sections of the windings are represented by circuit elements such as resistors, inductors, and capacitors [24], [32], [63]. The parameters can be calculated using the details of the transformer design using analytical equations [8] or FEM modeling of transformer windings [14]. Afterward, these parameters are changed to model various mechanical defects and to predict their effect on the FRA traces. Studying the effects of parameter changes helps to discover correlations between the frequency response variations and faults.

An example of one section of such models is shown in Figure 2-18. In this example, the windings are divided into several sections, and each section is then represented by different circuit parameters [56]. A description of each element is also given in Table 2-9. The white box model turned out to be more suitable for FRA interpretation studies as it can lead to a better understanding of how the physical dimension of transformer windings changes the features of corresponding FRA responses [64]. However, the white box model requires transformer design data, which may not be accessible in most cases by utilities but by original equipment manufacturers.

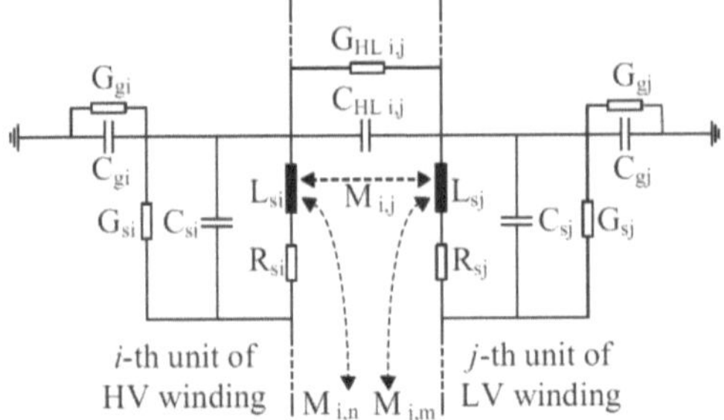

Figure 2-18: Lumped model of a two-winding transformer [8]

Table 2-9: Description of the circuit elements of the transformer model

Elements	Description
R_s	Series resistance of each section
L_s	Series inductance of each section
C_s	Capacitance between adjacent sections
C_g	Capacitance between each section and the grounded plate near it
G_g, G_s, G_{HL}	Conductance representing the dielectric loss of the paper-oil insulation system
C_{HL}	Capacitance between sections of LV and HV windings
M	Mutual inductance of each section to the sections of the same or the other winding

2.7.2.2 Black Box Model

The black box model is a mathematical model which is based on measurements on the transformer's external terminals [22], [23], [65], [66]. The model parameters are determined in a way that the simulated input-output response matches the measured data on the transformer terminals. Thus, the model parameters have no physical meaning or relationship with the transformer's geometry and cannot simulate the internal behavior of the winding.

In Ref [67], an advanced and modified black box model of a transformer is presented for the protection of load against lightning transients. The suggested model is based on black box, two ports four terminal network theory. For No load conditions, transformer parameters are calculated on two resonance frequencies of 450 kHz & 1 MHz using a Fast Fourier Transform. MATLAB/Simulink is used for simulation analysis. Both time and frequency domain analysis validate the accuracy. The resemblance between the measured and calculated results confirmed the precision of the proposed model when an impulse of 1.2/50µs is applied to transformer terminals.

2.7.2.3 Grey Box Model

The grey-box model is a compromise between the white-box model and the black-box model. Normally the circuit structure of a grey box has the same or similar circuit topology as the white-box model, but its parameters of the circuit are estimated from the given frequency response [68].

In 2002, J. Pleite, E. Olias, A. Barrado, A. Lazaro and J. Vazquez from Univ. Carlos III, Madrid, Spain proposed a method to construct the grey box model [69], [70]. The magnitude response of FRA data was used to demonstrate the grey box model. In the model, the number of resonance and anti-resonance points decides the number of the unit cell. Each cell dominates the response in a specific frequency sub-band and is assumed to have very limited influence at other bandwidths. However, this method does not have good repeatability. When it is applied to the artificial transformer equivalent winding circuits, the calculated values of the circuit elements may deviate from the original data.

In 2017, R. Aghmasheh and V. Rashtchi from the University of Zanjan, Iran, and E. Rahimpour from ABB, Germany put forward some new thoughts for the grey box modeling [71]. Weibull Distribution function is used in the estimation of inductance. An exponential function is used in the estimation of series resistance. Those ideas make the construction of a grey box model simple, fast, and robust. In [72] artificial method is used for grey box analysis of transformer parameter calculations. The numbers of the unknown parameter are reduced using both the Weibull Distribution function and the exponential function.

Application of Models

- When detailed data and the internal structure of the transformer are not present, the black-box model is appropriate to acquire the high-frequency behavior of the transformer.

- The white box models do not require measuring data and use transformer design documents and geometrical data to calculate the model parameters. Such models can describe the internal transient voltages with acceptable accuracy. In the White box model complexity is high; bandwidth is low as compared to grey and black box. But it allows a deeper system view. The model replicates the behavior of the transformer winding.
- The grey-box model lies in between the black-box and white-box model. The grey-box model is useful when there is no knowledge about the internal geometry and material properties available.

2.7.3 Intelligent Systems

Artificial intelligent (AI) methods comprise such methods as decision tree, genetic algorithm, fuzzy logic, neural networks, and so on. Some researchers use such methods to estimate the parameters of a transformer's high-frequency model, using data obtained from frequency-response measurements [73]. In other words, AI methods are employed to build a high-frequency model from real measurements. Some other researchers implement AI techniques for the recognition of frequency-response patterns of the winding to detect any failure of the internal winding insulation [74]. In this regard, the fault type is also determinable using frequency response classification through AI methods [75], [76], [77]. Some other works use AI methods for assessing the risk of power transformer failures [78].

In literature, many efforts have been made where machine learning (ML) methods are implemented for the detection and identification of transformer winding faults using FRA. Velasquez et al. (2011) [25] employed a decision tree model to classify low-frequency and high-frequency faults. Data from 500 transformers were employed for time-based comparison. Bigdeli et al. (2012) [19] employed a support vector machine (SVM) model to discriminate between different mechanical fault types. In this work, only two types of transformers were employed, a classic 20 kV transformer and a model transformer. Ghanizadeh et al. (2014) [32] employed an artificial neural network (ANN) to identify electrical and mechanical faults with good accuracy. However, a single 1.2 MVA transformer model was employed in this research. Gandhi et al. (2014) [26] used nine statistical indices and 90 FRA cases to develop a three-layer ANN. The main disadvantage of this work is that only a single transformer model was employed to generate the database. Aljohani et al. (2016) [20] employed digital image processing to automate fault identification; however, only two transformers of different ratings were simulated for this purpose. Luo et al. (2017) [19] employed ANN to recognize simulative winding deformation faults to different extents. In this contribution, transfer function, zeros, and poles were considered to be the features of ANN; however, the faults were simulated only in a transformer simulation model. Liu et al. (2019) [27] used the SVM model to diagnose three different types of faults, i.e., disk space variation, radial deformation, and inter-turn fault. Numerical indices were employed to extract eight features from FRA data. An accuracy rate of 96.3% was obtained in this contribution. Zhao et al. (2019) [30] introduced a winding deformation fault detection method. The method is based on the analysis of binary images extracted from FRA traces to improve FRA interpretation. The digital image processing technique is used to process the binary image and the outcome of this method is a fault detection indicator. Duan et al. (2019) [31] proposed a deep learning algorithm to detect inter-turn faults in transformers with an accuracy of 99.34%. However, the feasibility of this classification algorithm was assessed only with simulated and preliminary experimental data. Mao et al. (2020) [28] proposed a support vector machine (SVM) model to identify the winding type, which is critically important from an asset management point of view. However, with the changes in the transformer rating, the dominated frequency region also changes, which makes it difficult to generalize this model.

The results of these reports show the potential of AI and machine learning algorithms for fault diagnosis and classification. However, the performance of some classifiers, e.g., decision trees, can be further improved by using ensembled machine learning models. Most of these studies employed small datasets. Additionally, fixed-frequency sub-bands are considered to calculate the numerical indices, whereas standards state that the ranges of frequency sub-bands cannot be fixed as they depend on transformer ratings. Moreover, only a few faults are classified and a small number of transformers are employed, which decreases the diversity of fault patterns. Hence, a diverse dataset of various faults from the field is necessary to settle the criteria for using AI and ML methods.

2.8 Summary

In Table 2-10, a summary of the current state-of-the-art FRA interpretation methods is presented, where their pros and cons are listed.

Table 2-10: Summary of FRA interpretation methods

Methods	Pros.	Cons.
Numerical indices extracted directly from FRA traces	-Easy implementation -Quantitative method -No design information is required -Simple and easy to interpret - Possible to combine with AI for fault classification	-The results are only numbers hence lack of physical meaning -The few interpretation criteria available and are based on fixed frequency subbands -There are numerous indices proposed in the literature and a lack of a comprehensive comparison study to choose a few. -The few available interpretation criteria are not reliable, as many cases report their insignificance -The interpretation criteria are only valid for time-based comparisons
White box model	-Replicate the behavior of winding -Possible to develop a relationship between deviation in FRA trace and alteration of different parameters	-Design information is required -Bad accuracy at high frequencies (no evidence of the feasibility of modeling complex FRA responses of typical real power transformers) -Lack of a standard approach
Black box model	-Design information is not required	-Lack of a standard approach -Lack of criteria for interpretation
Grey box model	-Design information is not required	-Lack of a standard approach -Lack of criteria for interpretation
Intelligent systems	Possibility of fault detection and classification Possibility of developing an interpretation criterion	-A large database is required - Lack of a standard approach -Lack of knowledge of the best AI method

3 Transformer Failure Mode Analysis using FRA

In this chapter, a study on the characterization of the effects of failure modes on FRA is presented. The effect of different faults on FRA is objectively discussed through case studies from the field. Later, fault deviation patterns are generated that summarise the characteristic impact of individual faults on FRA results.

3.1 Features of an effective FRA assessment tool

Focusing on the current challenge to the existing assessment methods, an effective FRA assessment tool demands six tasks as appreciated in Figure 3-1. Task 1 deals with automatic detection and removal of noise, this is important because noise can be misinterpreted as faults, thus, identification of noise is necessary. Task 2 is related to the identification of low, medium, and high-frequency sub-bands, as the use of fixed frequency bands in existing assessment methods is a gross estimation that can be wrong in some cases, thus, a meaningful transformer-specific identification of sub-bands should be conceived.

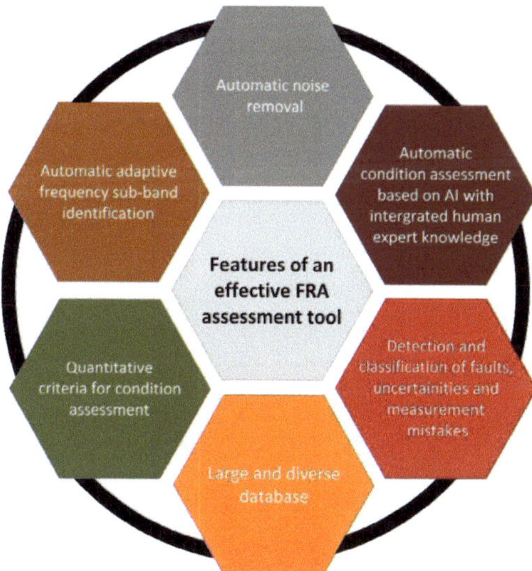

Figure 3-1: Features of an effective assessment tool

Task 3 deals with the quantitative criteria for the assessment of transformer condition based on FRA results. Task 4 is related to the collection of use cases to train the assessment tool with diverse datasets. Task 5 deals with the objective identification of fault type and discrimination of reproducibility issues and measurement mistakes. Finally, task 6 is the automatic condition assessment based on artificial intelligence methods with integrated human knowledge.

In the framework of this research, efforts are been made to develop and optimize the existing assessment methods to achieve an effective assessment tool which is described above.

3.2 Database

In the framework of this research, data from real case studies are collected from the field from different utilities, diagnosis companies, and working groups. The data is collected in collaboration with CIGRE working group A2.53 [13] (Advances in the interpretation of transformer Frequency Response Analysis (FRA)). The database collected in this study consists of 139 FRA results from 80 power transformers of different designs, ratings, and different manufacturers. Additionally, the FRA measurements, where winding deformations are simulated in distribution transformers are also used. The database comprises different types of transformers, i.e., generator step-up unit, transmission, distribution, shunt reactor, GIS connector, dry type, etc. The distribution of the power rating of transformers in the database is illustrated in Figure 3-2.

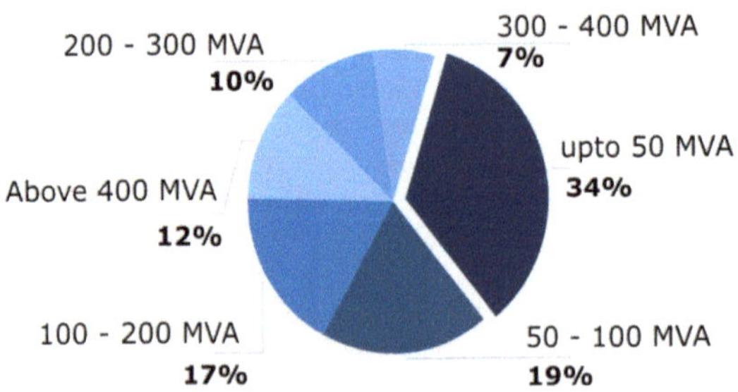

Figure 3-2: Content of database

3.3 Adaptive Frequency Division Algorithm

As stated in Chapter 2, the frequency response of a transformer has a fundamental relationship with the physical parts of the transformer. These physical parts dominate the frequency response in different frequency regions. Hence, by identifying these regions, different faults in power transformers can be classified. However, the ranges of these frequency regions or sub-bands depend upon many factors such as rating, size, core and winding structure, etc., and a general range cannot be concluded [5], [11].

The first approach for the development of an automatic frequency division algorithm is the identification and classification of different FRA patterns. In the database mainly, twelve features of FRA traces are recognized in different frequency sub-bands based on different transformer vector groups, winding structure, rating of the winding, etc. The classification of FRA traces based on different features in different frequency sub-bands is illustrated in Figure 3-3.

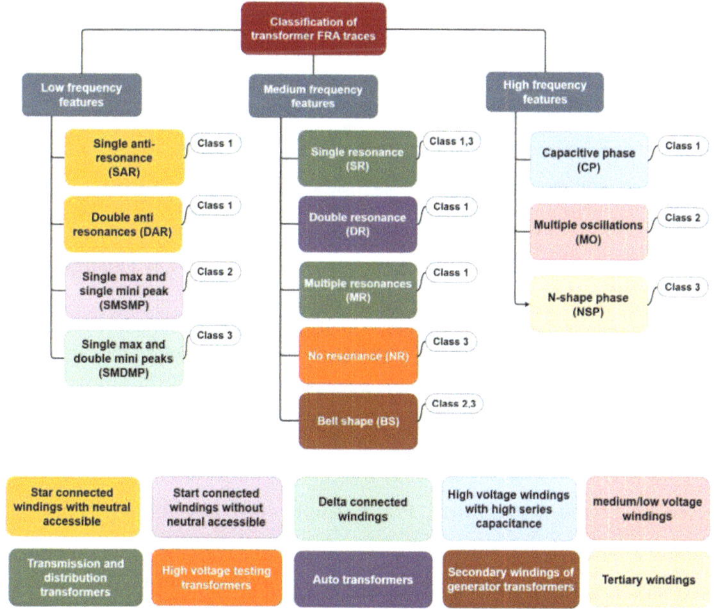

Figure 3-3: Classification of transformer FRA traces based on different features in the low, medium, and high-frequency range

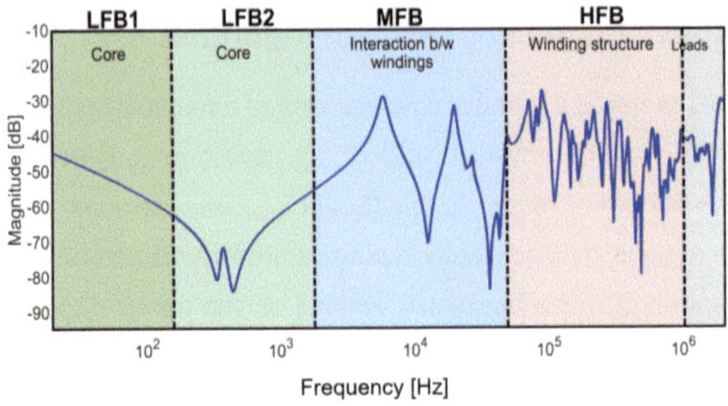

Figure 3-4: An example of Frequency sub-band division structure

Although there are at least 12 features of FRA traces, this does not imply that for the identification of each feature independent algorithms have to be developed. Instead, after analyzing all possible features it was concluded that it is possible to group them into fewer classes, thus, the 12 features are further classified into three classes based on the type of the transformer vector group and winding structure, etc. which are mainly responsible for the generation of different features are also mentioned in the bottom of Figure 3-3. Class 1 consists of FRA patterns belonging to star-connected windings with accessible neutral, autotransformers, and high voltage windings having high series capacitance disk windings. The FRA patterns of star-connected winding without neutral accessible transformers, medium and low voltage windings, and secondary windings of generator transformers belong to class 2, whereas, FRA patterns of delta connected primary and secondary windings, and ordinary disk or layer type windings, low impedance windings are grouped in class 3.

Based on these features an automatic frequency division algorithm is proposed which subdivides the entire frequency spectrum into four sub-bands, i.e., two low-frequency bands (LFB1 & LFB2), a medium frequency (MFB), and a high-frequency band (HFB). These frequency sub-bands are linked to different physical components of the transformer. For example, two low-frequency sub-bands (LFB1 and LFB2) are related to the core where magnetizing inductance (L_m) and equivalent network capacitance (C_{net}) dominate the response. Medium frequency sub-band (MFB) is dictated by the mutual inductances between windings (Mu) and inter-winding capacitances (C_w). The high-frequency region is controlled by the winding structure where a group of resonances caused by the winding inductance and series and ground capacitances. Figure 3-4 illustrates the 4 frequency sub-band structure for a transformer winding.

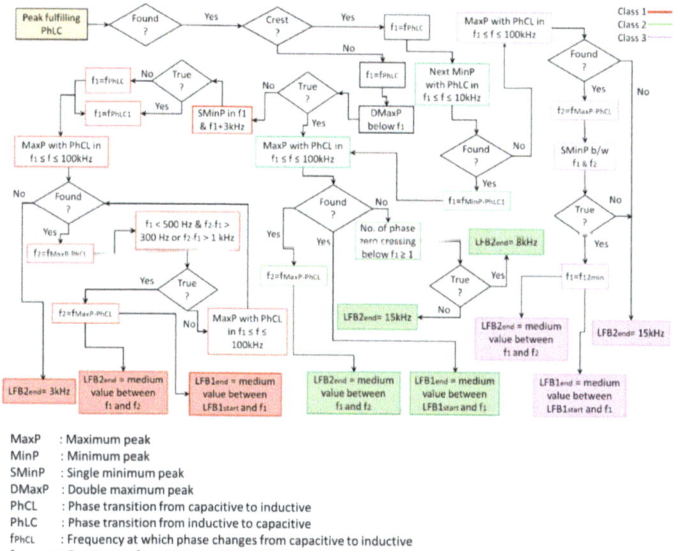

MaxP : Maximum peak
MinP : Minimum peak
SMinP : Single minimum peak
DMaxP : Double maximum peak
PhCL : Phase transition from capacitive to inductive
PhLC : Phase transition from inductive to capacitive
f_{PhCL} : Frequency at which phase changes from capacitive to inductive
$f_{MaxP\text{-}PhCL}$: Frequency of maximum peak at which phase changes from capacitive to inductive

Figure 3-5: Workflow of the algorithms for identification of the low-frequency sub-bands

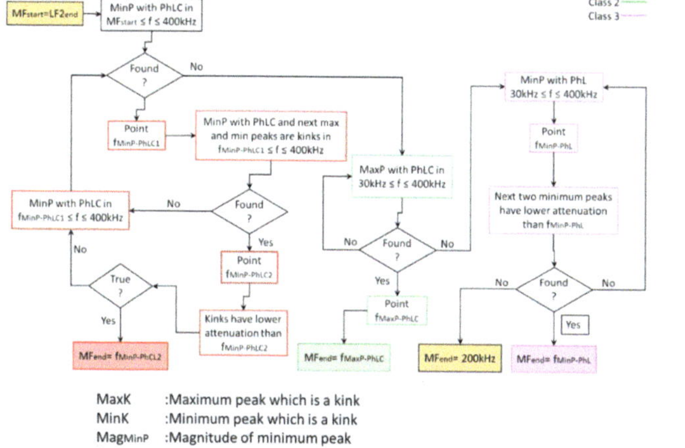

MaxK :Maximum peak which is a kink
MinK :Minimum peak which is a kink
Mag_{MinP} :Magnitude of minimum peak
$f_{MinP\text{-}PhLC}$:Frequency of minimum peak at which phase changes from inductive to capacitive
$f_{MinP\text{-}PhLC1}$:Frequency of minimum peak at which 1st phase transition from inductive to capacitive occurs
$f_{MinP\text{-}PhLC2}$:Frequency of minimum peak at which 2nd phase transition from inductive to capacitive occurs

Figure 3-6: Workflow of the algorithms for identification of the medium frequency sub-band

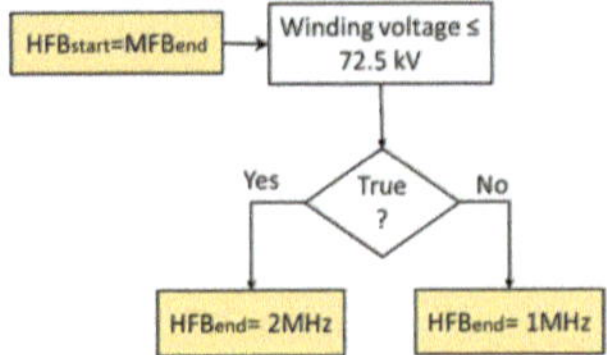

Figure 3-7: Decision workflow of the algorithm for identification of the high-frequency sub-band

It should be noted that this frequency division is based on the EE-OC transfer functions. The algorithms for the identification of LFB1, LFB2, MFB, and HFB are described in Figure 3-5, Figure 3-6, and Figure 3-7, respectively. The detailed procedures of these algorithms are discussed in the authors' previous work in [79].

3.3.1 Performance Evaluation

The performance of the adaptive frequency division algorithm has been evaluated by different case studies. The results are presented in Appendix B. It should be noted that the presented case studies include FRA plots that belong to different transformer windings, ratings, etc. The obtained results give evidence of the promising performance of the algorithms. The identified frequency sub-bands are in accordance with the proposed frequency sub-band structure. Thus, these algorithms provide the ability to automatically identify transformer-specific frequency sub-bands.

3.4 Study on the effects of failure modes on FRA

To develop and optimize FRA assessment methods, the characterization of the effects of different failure modes is necessary. In this section, the effects of different failure modes are discussed objectively to develop a failure assessment methodology.

3.4.1 Mechanical failure modes

As presented in Chapter 2, section 2.4.1, there are various forms of mechanical deformations. The effect of the six most common mechanical failure modes (clamping failure, bulk winding movement, conductor tilting, lead deformation, axial collapse, and buckling) on FRA are presented in the following:

3.4.1.1 Axial instability after clamping failure (AI-M)

A real case study of axial instability observed in a 3-phase, 240 MVA, 400/132 kV autotransformer [13]. The unit was switched out of service for investigation after a Buchholz alarm. A time-based comparison of FRA measurements on the common winding of phase A, before and after the fault is shown in Figure 3-8.

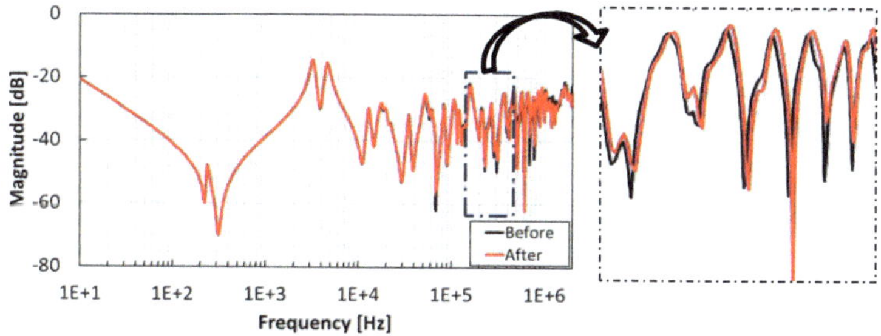

Figure 3-8: TF of A-phase LV winding before and after axial collapse

Figure 3-9: The axial collapse of A-phase LV winding due to clamping failure

The visual analysis of the FRA results shows some deviations and shifts of resonances in the high-frequency sub-band (HFB) where the response is dominated by the winding structure. At high frequency, the resonances are shifted to the right which can be attributed to reductions in series capacitances (C_s). As Cs reduces due to the reduction of the permittivity of pressboard spacers since according to previous investigations the permittivity is dependent on the applied pressure (density). The TF remains unchanged in LFB1, LFB2, and MFB. Moreover, the effect of this fault was not observed in other phases as the FRA measurements of the other phases were perfectly aligned with each other. After strip-down irreparable damage such as axial instability after clamping failure and twisting to A-phase LV winding was found as shown in Figure 3-9.

3.4.1.2 Bulk winding movement/axial displacement (TS-M)

In this failure the individual winding moves relative to the other winding. A real case study showing the effects of telescoping in the FRA plots was found in a single-phase, 21.6 MVA, 166/10 kV transformer [13]. The FRA measurements were performed and compared with the previous FRA measurements. Definite resonance frequency shifts for the response of LV winding were identified. During the subsequent strip-down inspection, there was evidence of damage due to the axial movement of the LV winding as shown in Figure 3-10.

Figure 3-10: The axial collapse of A-phase LV winding due to clamping failure

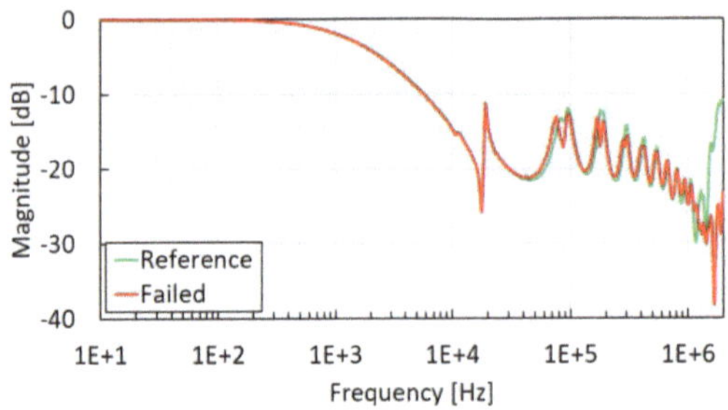

Figure 3-11: TF of A-phase LV winding before and after axial collapse

End-to-end short circuit TFs of LV winding before and after the fault event are shown in Figure 3-11. In low-frequency sub-bands (LFB1 and LFB2) the deviations between TFs are due to the different remanence conditions of the core. The TF in the MFB is unaffected. While in the HFB the TF of the deformed winding showed a significant shift of the resonances to the left. This can be attributed to the increase of ground capacitance (C_g) and/or series capacitance (C_s), as in axial displacement fault the winding is stretched or "telescoped" and then tightens due to a reduction in the radius of the winding, thus, C_g increase.

3.4.1.3 Conductor tilting (CT-M)

A real case study showing the effect of conductor tilting in the C-phase of a three-phase, 120 MVA, 275/33 kV, YNd transformer [13]. In this case, the time-based comparison was not possible, thus, phase-to-phase and sister unit comparisons were performed. The sister unit was measured with oil and the failed unit was measured without oil. During stripping, the tap winding of the C-phase was found with conductors tilting, as shown in Figure 3-12.

Figure 3-12: Conductor tilting of C-phase tap leads

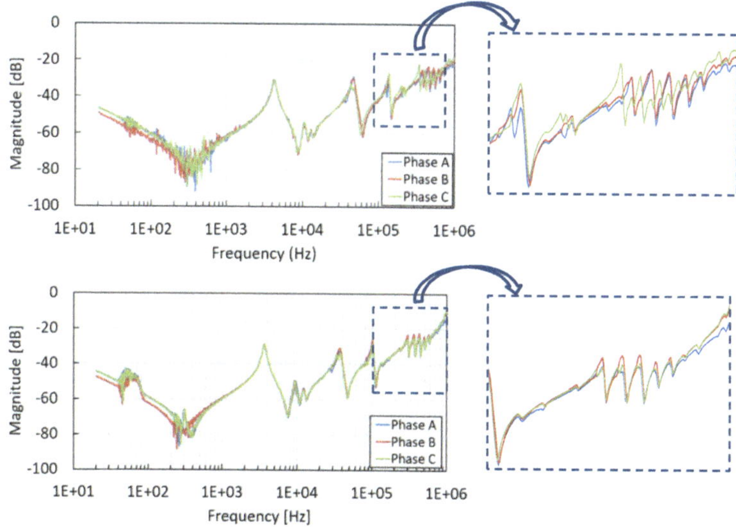

Figure 3-13: OC-EE FRA for failed unit (Top) and sister unit (bottom)

The FRA measurements on HV winding are shown in Figure 3-13. In the case of the sister unit, all three phases demonstrate a good match. While changes of resonance frequencies in the range related to the winding structure are evidenced on the C-phase of the failed unit. This can be attributed to the increase in the C_s of the windings. As can be appreciated that the deviations in the low-frequency subbands are due to the external interference and normal asymmetry of lateral and central phases.

3.4.1.4 Lead deformation (LD-M)

A case study of lead deformation after a short circuit test is reported. In this case, the unit is a 3-phase, 250 MVA, 400/155 kV autotransformer [13]. After the short-circuit event, the unit was internally inspected and a slight displacement and twisting of leads were found on the C-phase tap winding as shown in Figure 3-14.

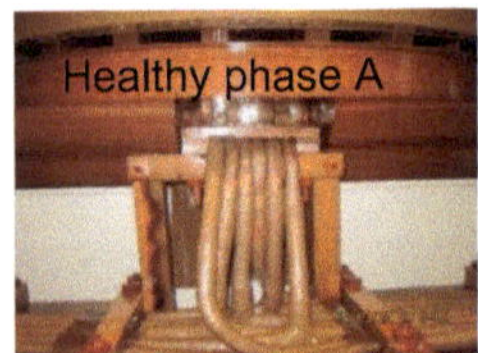

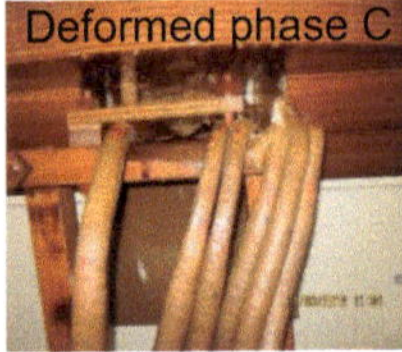

Figure 3-14: Normal leads phase A (left) and displaced/twisted leads phase C (right)

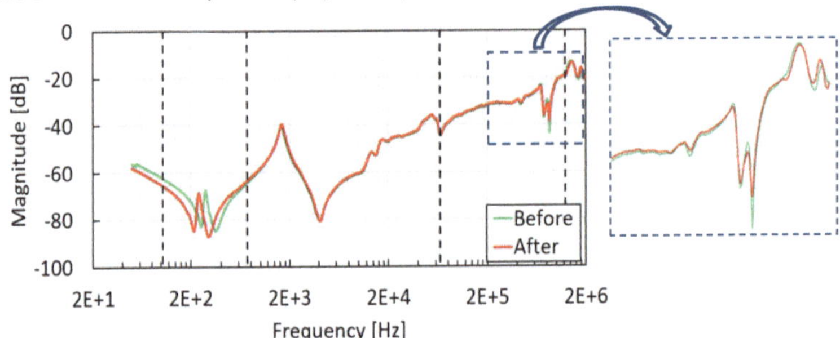

Figure 3-15: EE-OC TFs of C-phase before and after short circuit event

EE-OC TFs of HV winding before and after the short-circuit event are shown in Figure 3-15. The deviations in the low-frequency regions are due to a different core magnetization. The TF of the deformed winding showed a slight variation around 1 MHz, indicating slight changes in the series capacitance of the winding. It is important to mention that such mechanical deformation does not have a specific pattern, the deviations in the TFs depend upon the severity of the fault. However, from the presented case study, it is clear that such faults affect the FRA at high frequency.

3.4.1.5 Hoop Buckling (HB-M)

A real case study showing the effect of hoop buckling in LV winding of a single-phase, 600 MVA, 22/432 kV transformer [13]. The assessment is made by comparing measurements on sister units composing the bank of single-phase transformers. Figure 3-16 shows the LV winding EE-OC TFs of two identical transformers. The deviations in the LFB1 and LFB2 are due to the different core remanence. In MFB, the resonance points are slightly shifted to the left. While the main deviations lie in the HFB. In HFB, the response is shifted to the left, indicating an increase in the leakage reactance due to hoop buckling. The internal inspection showed significant hoop buckling of the inner LV winding as shown in Figure 3-16.

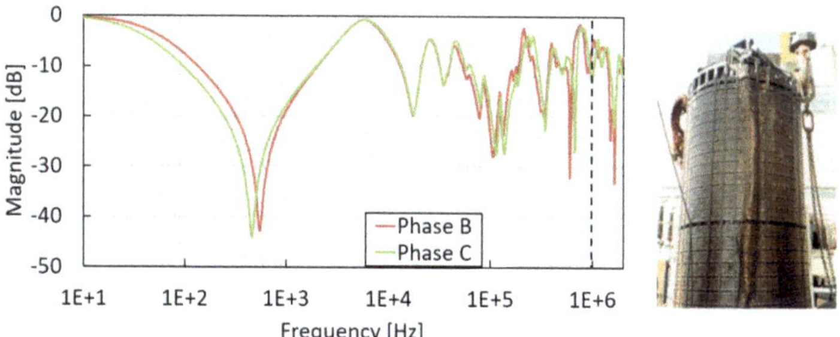

Figure 3-16: LV winding EE-OC TFs of two identical transformers (left) and hoop buckling fault is visible during internal inspection (right)

3.4.2 Electrical failure modes

As presented in Chapter 2, section 2.4.2 there are various types of electrical faults. The effect of the five most common electrical failure modes (clamping failure, bulk winding movement, conductor tilting, axial collapse, and buckling) on FRA are presented in the following:

3.4.2.1 Shorted turn fault (SCC-E)

A short circuit between turns is the most common electrical failure mode in transformer windings. Depending upon the severity, it can influence the characteristics of magnetizing inductance, winding resistance, and self-inductance of the windings. Shorted turn fault gives rise to large circulating currents, which leads to localized thermal overloading, thereby causing hot spots [35]. When a SCC-E fault occurs, a large amount of flux passes through the air instead of the core and surrounds the shorted-turn. Thus, the leakage flux is essentially increased at the fault location. Consequently, the magnetic reluctance and self-inductance of the corresponding limb changes. Accordingly, a case study is presented where a shorted turn fault at the U-phase occurred due to a lightning surge. Phase-to-phase comparison of EE-OC FRA measurements is presented in Figure 3-17. As can be appreciated the main deviation lies in the low-frequency sub-bands where the response of the core dominates.

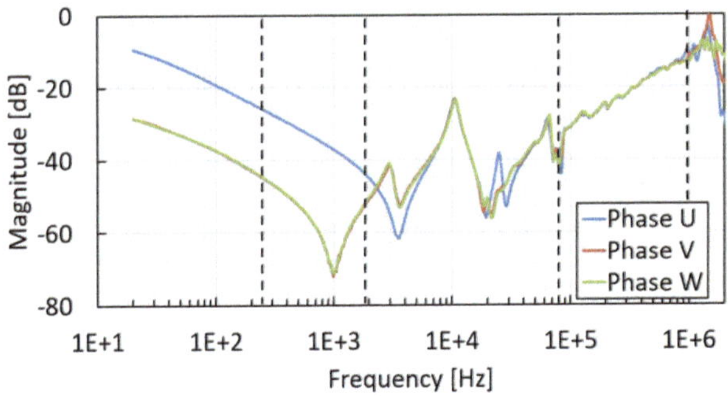

Figure 3-17: HV winding EE-OC TFs

3.4.2.2 Open circuit fault (OC-E)

An open circuit can be caused by connections that become loose or coils that become burned due to a catastrophic thermal failure. This results in very high impedances being inserted into the measurement circuit. The main indication of this fault is an increase in the attenuation with a vertical shift downwards. For complete open circuits, the results will often be lost in the noise floor of the measurement.

An illustration of a partial OC-E fault in a real case study is reported. The unit is a three-phase, 34 MVA 237/5.65 kV YNd11 transformer [13]. High voltage EE-OC FRA measurements are shown in Figure 3-18. It can be seen, that at low frequency, the attenuation is increased and the primary core resonance shape changes to account for the faulty winding. At high frequency (where the structure of the winding dominates the response) the resonance points do not experience a significant shift. This indicates that OC-E fault does not necessarily change the reactance of the winding, as an alternative, the resistance significantly changes which produces vertical shifts in the FRA trace.

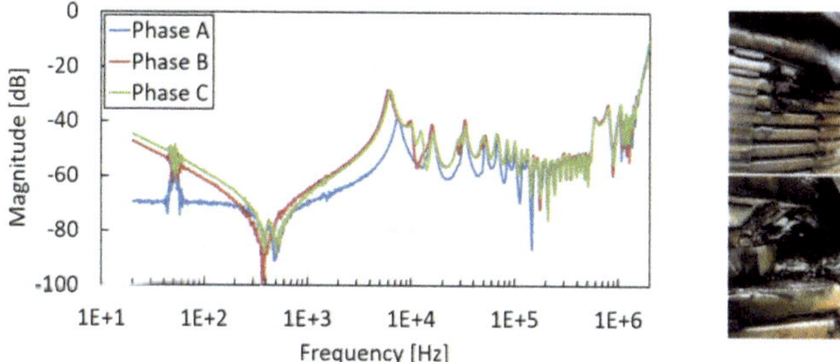

Figure 3-18: HV winding EE-OC TFs (left) and internal inspection of OC-E fault (right)

3.4.2.3 Core ground loss (CGL-E)

In normal operation, transformer cores are tied to ground potential at one point. This single point earthing can be lost during transportation/service conditions. Generally, the missing core ground fault changes the LV winding-to-ground capacitance, which results in the shifting of the HV winding response in the medium and high-frequency range. This can be explained, using the circuit in Figure 3-19, where a simplified equivalent circuit of a two-winding transformer with and without core ground is shown. Under normal conditions, when the frequency response of the HV winding is measured by applying the voltage at the terminal of the HV winding, the voltage will be induced in the LV winding, which is electromagnetically coupled with the HV winding. Consequently, at the measuring impedance, in addition to the HV winding current, there is a capacitive current flowing from the LV winding. In the case of a missing core ground, the coupling between the LV winding and ground is reduced, increasing the potential of the floating LV winding. This additionally increases the capacitive current flowing from LV winding through the measuring impedance. It will change the frequency response of the HV winding in the middle and high-frequency regions.

A real case study of core ground loss was reported in a three-phase 140 MVA 353/13.8kV YNd5. In some transformers like this, the connection to the ground of the core is made externally. This makes it easier to detect and investigate the effect of core grounding-related problems. The FRA traces are measured with grounded core and floating core conditions as shown in Figure 3-20. The effect of CGL-E fault is most obvious in the MFB and HFB. As mentioned earlier, CGL-E fault results in additional capacitive current flowing from LV winding and inter-winding capacitances towards the measuring impedance. This capacitive component will change the TF of the HV winding in MFB and HFB.

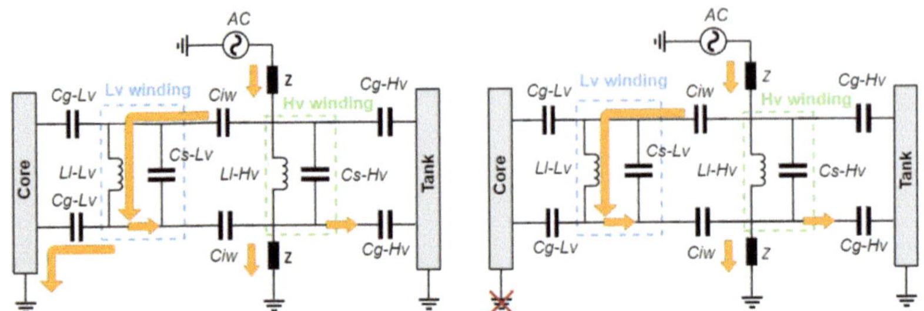

Figure 3-19: Equivalent circuit of a two-winding transformer for diagnosis of core ground loss fault; (left) grounded core, (right) core ground loss

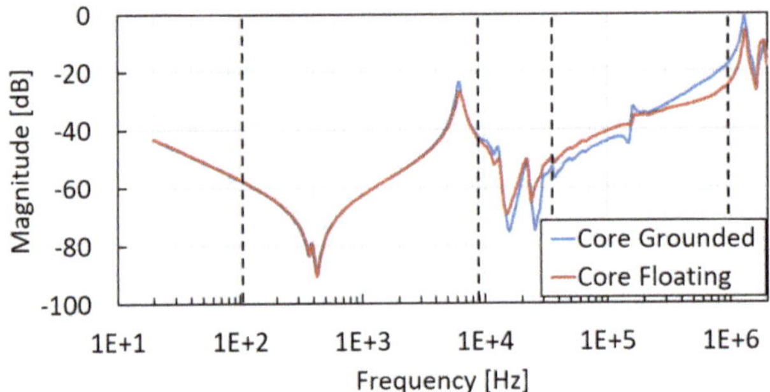

Figure 3-20: HV winding EE-OC TFs for grounded and floated core

3.4.3 Summary of deviation patterns of failure modes

In summary, the effects of different failure modes on FRA can be characterized by certain deviation patterns. Table 3-1 shows the summary of deviation patterns obtained from the presented case studies. The deviation patterns are characterized according to the frequency sub-bands. In Table 3-1, the sub-bands of the FRA plot in which the effect of failure modes can be appreciated are marked by three symbols, i.e., no deviation (●), slight deviation (▲), and obvious deviation (■).

Table 3-1: Summary of deviation patterns of failure modes

Failure modes	Frequency sub-bands				Comment
	LFB1	LFB2	MFB	HFB	
Clamping failure (CF-M)	●	●	●	■	-Deviations at high frequency with a shift of several resonance points and or produce new peaks and valleys
bulk winding movement (BWM-M)	●	●	▲	■	-Frequency shifts at medium and high frequency (typically shift to the left)
Conductor tilting (CT-M)	●	●	▲	■	-Slight frequency shifts (typically to the left). Difficult to identify small deformations
lead deformation (LD-M)	●	●	●	▲	-Typical pattern is difficult to identify
Telescoping (TS-M)	●	●	▲	■	- Newly created resonance peaks or valleys at high frequency are the key indicators
Buckling (BL-M)	●	●	▲	■	- It can shift or produce new resonance peaks and valleys at medium and high frequency
Shorted turn (SCC-E)	■	■	●	●	-The EE-OC FRA trace assumes a similar behavior as the EE-SC test with reduced attenuation at low frequency
Open circuit (OC-E)	■	■	■	■	-Increase attenuation in the entire frequency spectrum or capacitive effect or response is lost in the noise floor of the instrument
Core ground loss (CGL-E)	●	●	■	■	- Decrease attenuation at high frequency

3.4.4 Factors Affecting Reproducibility

For FRA measurements the difference between traces does not always belong to the abnormal transformer, in fact sometimes measurement mistakes introduce differences in various frequency sub-bands. Small differences in certain frequency sub-bands might be produced by different factors, which hinders fault diagnosis and may lead to the wrong interpretation of FRA results. If the effect of different factors on FRA is known then these factors can be eliminated during interpretation and attention can be turned to the transformer. As appreciated in Figure 3-21, the factors affecting reproducibility can be divided into three groups, i.e., transformer conditional factors, measurement-related factors, and external factors. Following examples from real case studies are given to investigate their effect on FRA measurements.

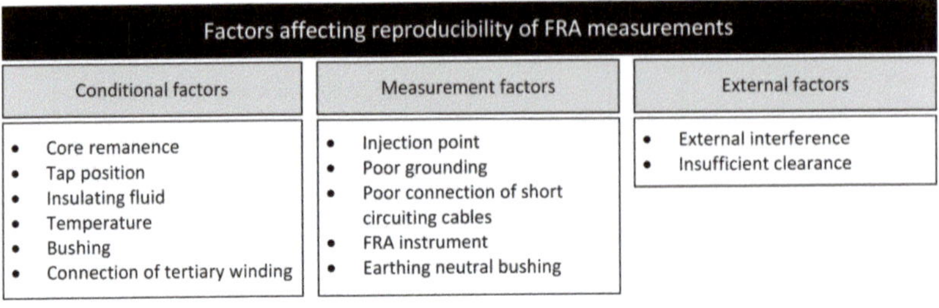

Figure 3-21: Summary of the factors affecting reproducibility of FRA measurements

3.4.4.1 Transformer condition-related factors

3.4.4.1.1 Residual magnetization (RM-R)

Residual magnetization is a frequently occurring phenomenon caused when a transformer is subjected to electrical excitation. When excitation is removed some magnetization is left behind in the core that changes the magnetic state of the core. Residual magnetization is the flux density that remains in the core steel, thus, changing magnetizing inductance. These changes become apparent in the low-frequency range, where they create a characteristic shift of the trace to the right. The other frequency sub-bands remain unaffected by this phenomenon. An example of a residual magnetization case study is shown in Figure 3-22.

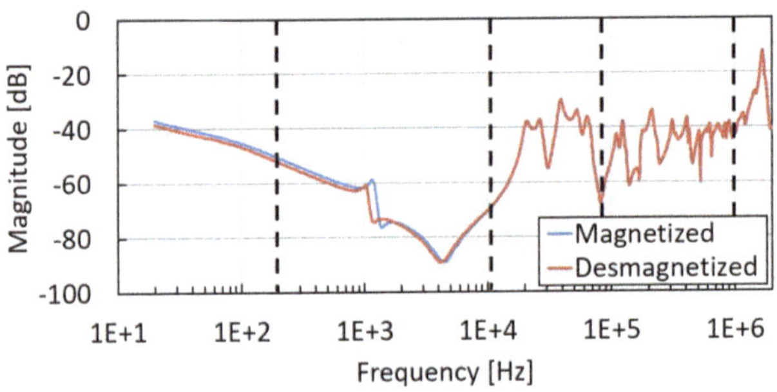

Figure 3-22: Effect of residual magnetization on EE-OC FRA measurement

As discussed before the main characteristic of this factor is the reduction in magnetizing inductance and shift of trace to the left in the LFB1 and LFB2 in EE-OC

TFs. As can be appreciated, the effect of residual magnetism can be simply identified and usually does not influence further analysis.

3.4.4.1.2 Tap position (TP-R)

According to the survey conducted by CIGRE WG A2.37, tap changer failures account for 31% of failure locations in substation transformers [34]. This failure could be either in the transformer tap changer selector switch or it can be in the diverter switch. Thus, it is beneficial to keep a record of FRA signatures for all taps during the commissioning of the transformer. These recorded FRA data will become useful during transformer diagnosis. In IEEE and IEC standards, it is recommended to perform the FRA test on extreme raise tap position.

Transformer frequency response is highly influenced by the tap position in the low, medium, and high-frequency regions of the FRA spectrum. With tap position variation the number of winding-turns will be simply changed. Therefore, the self and mutual inductance, as well as the series capacitance, is considerably changed, and eventually, frequency response is changed. Moreover, tap position variation will change the winding resistance from its nominal value in the nominal tap position. Thus, the influence of winding resistance variations may become noticeable in low as well as high-frequency regions of the FRA spectrum.

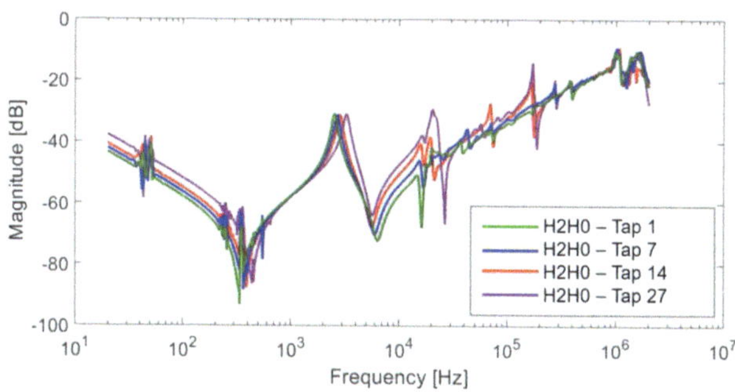

Figure 3-23: Effect of Tap position on EE-OC FRA measurement

A 300 MVA 405/115/21 kV, transformer with linear tap-changer arrangement in HV winding, has been examined. The influence of different tap positions on the HV side is investigated as depicted in Figure 3-23. It is confirmed that the position of the tap changer switch greatly influences the characteristics of the FRA signature. Moreover, for some transformer designs, the direction from which OLTC returns to the neutral position also impacts the frequency response. The IEC standard [11] indicates that the direction of movement shall be in the lowering voltage direction unless otherwise specified. Thus, it is recommended to keep a record during measurement to prevent any confusion in FRA interpretation.

3.4.4.1.3 Insulating fluid (IF-R)

Typically, power transformers are transported without oil, and fingerprints FRA are measured at the factory with the presence of oil. Moreover, the application of different insulating fluids at factories and fields is also a common practice. Thus, it is important to investigate the effect of oil on FRA measurements and it should be carefully reported so that it cannot be misinterpreted as faults.

Generally, the absence of oil decreases the dielectric constant and decreases capacitance, thus shifting all resonant points of the entire frequency response trace to higher frequencies. Similarly, an insulating liquid with a dielectric constant higher than that of oil (e.g., natural ester) will shift the response towards lower frequencies. A case study showing the effect of oil on FRA is shown in Figure 3-24. As can be appreciated the absence of oil shifts the FRA trace to the right in the entire frequency spectrum.

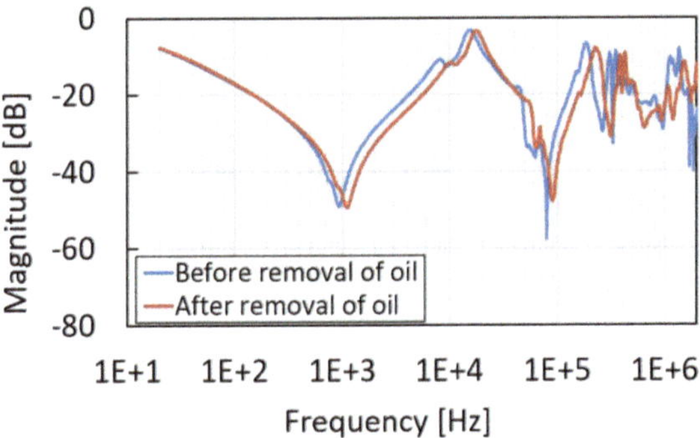

Figure 3-24: EE-OC TFs with and without oil (10 MVA 63/6.6 kV)

Another case study is reported in Figure 3-25, where factory tests, "fingerprints" were performed with natural ester and field tests were conducted with mineral oil in the transformer tank. As can be appreciated natural ester which possesses a higher dielectric constant shifts the response to the left in the entire spectrum.

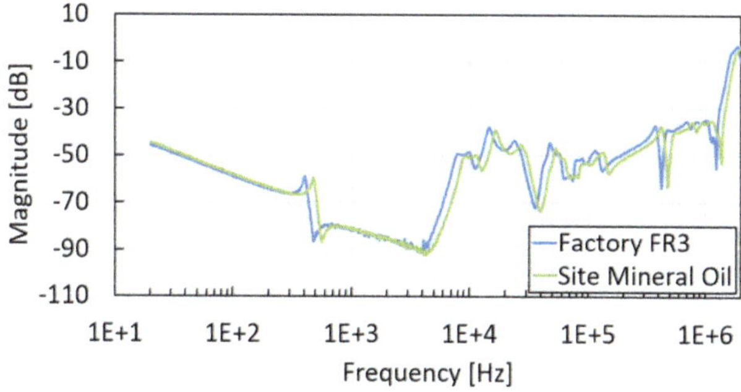

Figure 3-25: Effect of different insulating fluids (70 MVA 138/13.8 kV)

3.4.4.1.4 Temperature (TEMP-R)

Usually, factory measurements are performed at room temperature. However, field measurements are carried out at a variety of temperatures, from ambient to near-to-service temperature. Moreover, a transformer is subjected to variable temperature during its lifetime. Thus, the effect of temperature is an important concern for repeatable FRA measurements. Generally, due to the temperature rise of oil, the permittivity of oil-impregnated insulation papers and press boards varies which may affect the resistance and/or series and parallel capacitances. The extent of influence depends on the difference in temperature between the measurements.

A case study to explain the effect of temperature on FRA is presented in Figure 3-26. In the example, two FRA measurements are shown which are measured at the factory (red) at room temperature 20°C and field (blue) at -20°C. Both measurements were carried out without oil. It can be seen that the temperature increase caused a shift of resonance frequencies in the entire frequency spectrum.

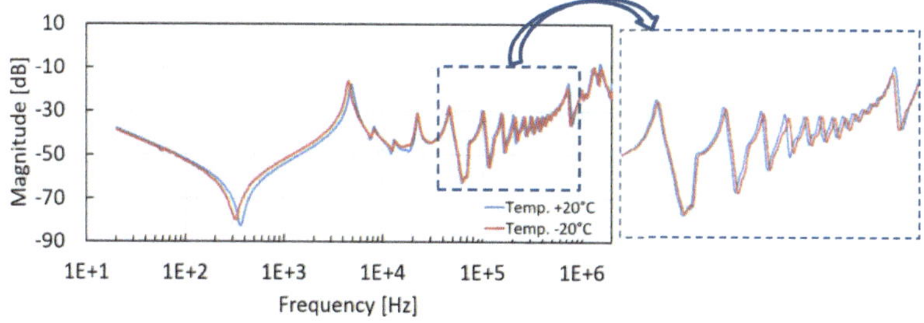

Figure 3-26: Effect of temperature (170 MVA 735/315/12 kV)

3.4.4.1.5 Bushing (BUSH-R)

Sometimes factory and field measurements are carried with different bushings or during the service life of the transformer, the bushings are replaced. Thus, the effect of the bushing on FRA is an important concern. Generally, there are two important characteristics of the bushings that can affect FRA responses, i.e., the capacitance of the bushing to the ground and the length of the bushing. The capacitance of the bushing to the ground changes the net parallel capacitances of the circuit. The length of the bushing could influence the length of the grounding braid cable. As for larger bushing much longer, grounding braid cables are required which include extra inductance in the circuit that can affect the frequency response.

A real case study is reported where different bushings are applied during factory and onsite testing. Figure 3-27 shows the comparison of TFs measured with different bushings. As can be seen, the effect of different bushings is appreciated mainly in the high-frequency region.

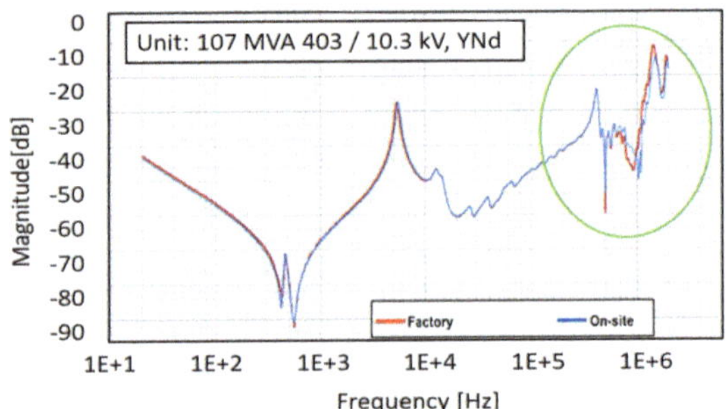

Figure 3-27: Effect of different bushings during factory and on-site FRA measurements (107 MVA 403/10.3 kV) [13]

3.4.4.2 Measurement-Related Factors

3.4.4.2.1 Reverse injection (RI-R)

When transformer design requires a non-uniform insulation distribution along the windings, the measurement results obtained by injecting the measurement signal to different ends of the winding will differ. Thus, information about the injection point (phase injection or neutral injection) should be carefully reported during the FRA test. To discuss the effect of injection point on FRA a case study is reported.

The unit is a 10 MVA, 64.5/6.9 kV transformer where each end of the winding was designed for a different insulation level. As a result, the neutral end of the winding had a smaller bushing and a different capacitance to earth than the one on the line end. Figure 3-28 shows the EE-OC measurements performed at two different injection points. Due to the non-uniform distribution of insulation a deviation between TFs at the high frequency can be observed. It is therefore recommended to use the same direction of measurement for analysis and keep a record of the injection point.

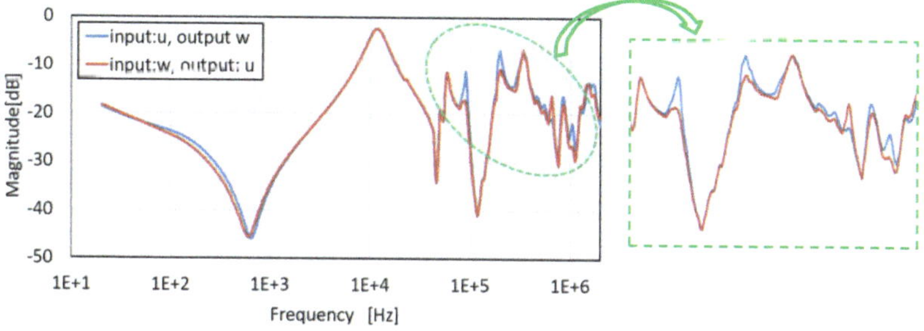

Figure 3-28: Influence of reverse injection (10 MVA 64.5/6.9 kV)

3.4.4.2.2 Poor grounding practice (PGP-R)

The braid lead which is used to ground the screen of the high-frequency cable at the base of large HV bushings is one of the key parameters limiting the repeatability of FRA measurement. To achieve good repeatability, it is recommended to keep the braid leads as short as possible as shown in Figure 3-29. The braid leads mainly have inductive characteristics and when its length is increased the inductance of the grounding path increases, which influences the high-frequency portion of the FRA trace. An example of the influence created by poor grounding practice with increased braid length (twice in length) is shown in Figure 3-30. The low and medium frequency sub-bands are not affected by the poor grounding practice. The influence of poor grounding practice exists at high-frequency generally above 100 kHz. The effect gets more prominent when the connection between the braid lead and the bushing flange is lost due to some poor connection or due to some insulation coating available on the bushing flange. An example of the influence, created by removing the braid leads during measurements, is shown in Figure 3-30. The result shows that poor grounding practices significantly affect the characteristics of the measured TF mainly at high frequency. Based on these measurements, the effect of poor grounding can be easily identified in the high-frequency region, and it should be eliminated before suspecting the transformer.

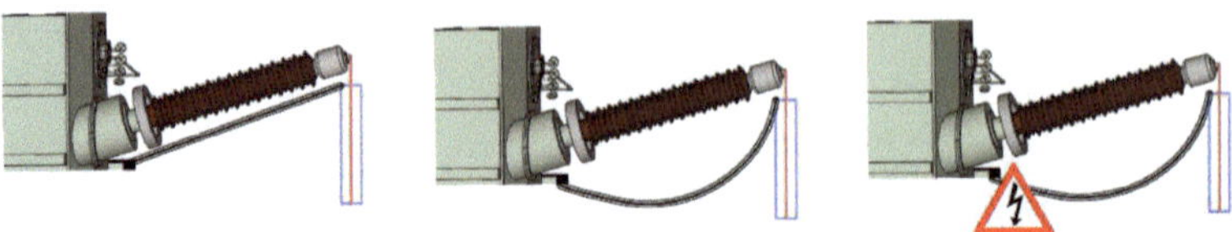

Figure 3-29: Grounding of braid cable (10 MVA 64.5/6.9 kV); (left) proper grounding, (center) poor grounding, and (right) loss of ground connection

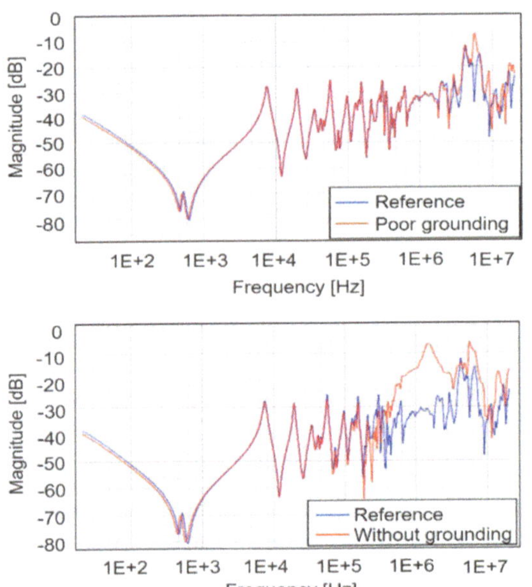

Figure 3-30: Influence of ground practice (12 MVA 110/10.6 kV); (top) poor grounding, and (bottom) loss of grounding connection

3.4.4.2.3 Poor connection of short-circuiting cables (PCSC-R)

The end-to-end short circuit measurement (EE-SC) removes the influence of core properties. It is performed from one end of an HV winding to another while the associated LV winding is shorted. To achieve good repeatability, it is recommended that all LV windings are shorted on 3φ transformers to create a three-phase equivalent short circuit. This ensures all three phases are similarly shorted to give consistent impedance. The loose connection of short-circuiting jumpers can introduce impedance to the path and can be detected by FRA. The influence is implemented in an example in which all three phases of LV are shorted through resistors. Figure 3-31 shows a short circuit measurement performed from the HV side. The comparison is made between good and poor connections (through resistors) of short-circuiting jumpers on the LV side. It can be seen that the effect is only at very low frequency which can be easily identified, and should be eliminated during fault diagnosis.

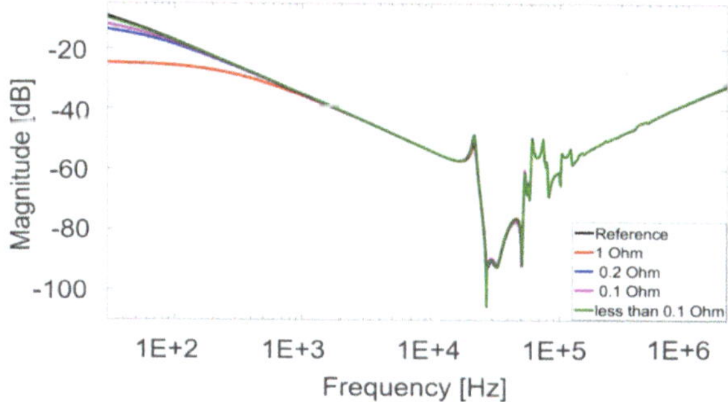

Figure 3-31: Influence of short circuit jumpers on EE-SC FRA (12 MVA 110/10.6 kV)

3.4.4.2.4 Measurement instrument (MI-R)

In FRA, different measurement instruments can be employed for factory and field measurements. The expected lifetime of power transformers is much larger than the measurement instruments. Thus, in comparing the field measurements with factory measurements which were carried out several tens of years before, the employed measuring instruments would be different. Therefore, the effect of several currently available instruments on FRA is important to investigate.

To discuss the effect of measuring instruments on FRA, three commercially available instruments are employed to measure FRA signatures of the same transformer, i.e., FRAnalyzer [80], FRAX [81] and FRANEO [33]. The attached cables to all three instruments follow Method I of IEC60076-18 (2012) [11]. The unit employed in this research is a three-phase, 12 MVA 126.2/10.6 kV autotransformer. Figure 3-32 shows the comparison of FRA performed with three different measuring instruments. The comparison results show that the influence of the difference in the measuring instruments is negligible in practical diagnoses.

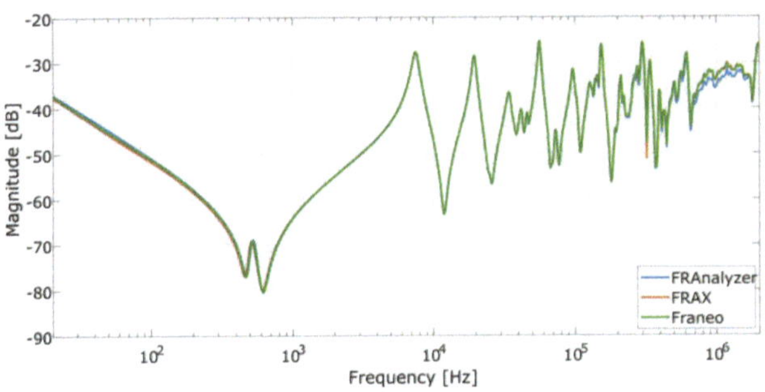

Figure 3-32: Influence of measurement instrument on EE-OC FRA (12 MVA 110/10.6 kV)

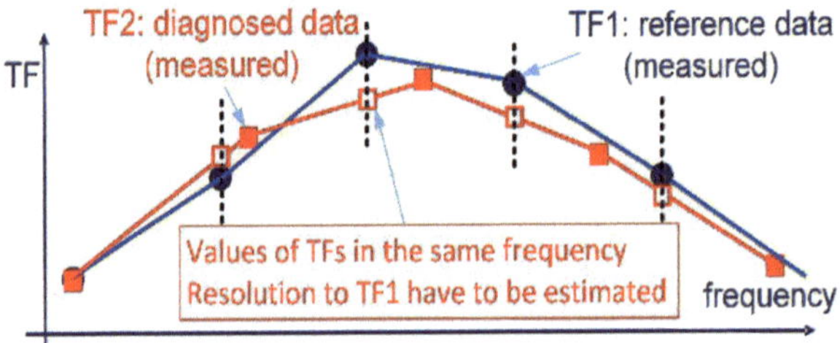

Figure 3-33: Calculation of numerical index of two TFs with different frequency resolution [82]

Another possible issue regarding the application of different measuring instruments is the different frequency resolutions. Different instruments possess different frequency resolution settings because in the international standards, IEC60076-18:2012, the frequency resolution is not specified in detail, and in practice, it depends on the instruments employed in the measurements. For fault diagnosis using numerical indices, the frequency resolutions in the measurements have to be identical. When the frequency resolutions are not identical, TF values with the same frequency resolution have to be estimated, for example, by employing an adequate interpolation method as shown in Figure 3-33. For this purpose, a comparison method adopting cubic spline interpolation (CSI) was proposed to estimate TF values in the same frequency resolution as the reference data. In author's previous work [83], a case study is presented where three commercially available instruments are employed having different frequency resolutions. The TFs were compared by employing the CSI method. The comparison results validate the previous findings that the influence of employing a different measuring instrument on the repeatability of measurements in FRA is negligible in practical applications when the connection methods for the measuring leads are the same. The detailed results can be found in authors publications [82], [83].

3.4.4.3 External factors

3.4.4.3.1 Noise (case 47 and 50) (N-R)

The FRA measurements performed in substation environments comprise a variety of noise that can influence the FRA signatures. Mainly there are two types of noise sources, i.e., narrowband noise and broadband noise. A narrowband noise is due to the power frequency noise and its harmonics. Typically, the narrowband noise affects the FRA plots in the low-frequency area between 30 Hz and 100 Hz. whereas, the broadband noise is caused by the noise bed of the connected FRA measurement equipment. The dynamic range defined the noise bed of an FRA instrument. IEC 60076-18 standard defines the minimum dynamic range for a device: -90 dB to +10 dB [11]. Therefore, it is recommended to select an FRA test instrument with a dynamic range greater than 100 dB. These noise effects only take place in the very low-frequency area where the linear magnetization inductance dominates the overall frequency response. The asset type and size of the transformer windings are also crucial elements for influencing the frequency response results. As the signal attenuation increases with the inductance and, thus, the response signal becomes more sensitive to noise. Figure 3-34, shows the effect of both types of noise. The effect of both types of noise appears at low frequency (before the first resonance point in the end-to-end open circuit FRA signature) causing several small peaks and valleys.

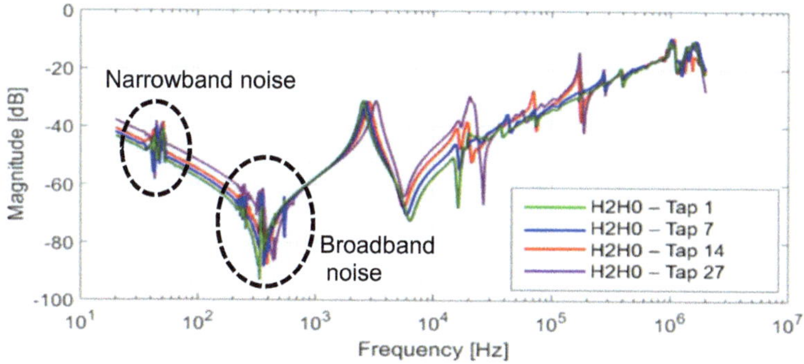

Figure 3-34: Effect of noise on EE-OC FRA measurements

In general, the interpretation of the mechanical failure modes of the power transformer is not influenced by the presence of noise. In contrast, however, the interpretation of electrical failure modes of a power transformer such as shorted turn fault, open circuit fault, and shorted core laminations can be influenced by noise. In Chapter 7, section 7.2.2 an algorithm is presented for the detection and removal of noise in FRA measurements.

3.4.5 Summary of deviation patterns of factors affecting reproducibility

In summary, the effects of different factors on FRA can be characterized by certain deviation patterns. Table 3-2 shows the summary of deviation patterns obtained from the presented case studies. The deviation patterns are characterized according to the deviations that appeared in different frequency sub-bands. In Table 3-2, the sub-bands of the FRA plot in which the effect of factors can be appreciated are marked by three symbols, i.e., no deviation (●), slight deviation (▲), and obvious deviation (■).

Table 3-2: Summary of deviation patterns of factors affecting reproducibility

Factors	Frequency sub-bands				Comments
	LFB1	LFB2	MFB	HFB	
Residual magnetization (RM-R)	▲	▲	●	●	-Deviations at low frequency.
Tap position (TP-R)	■	■	■	■	-Deviations in the entire frequency spectrum depending upon the position of the tap
Tap position direction (TPD-R)	●	●	▲	●	-Typical pattern is difficult to identify
Insulating fluid (IF-R)	▲	▲	▲	▲	-Entire frequency spectrum shifts
Temperature (TEMP-R)	▲	▲	▲	▲	-Entire frequency spectrum shifts
Bushing (BUSH-R)	●	●	●	■	-Typical pattern is difficult to identify
Reverse Injection (RI-R)	●	●	●	■	-Deviations at high frequency
Poor grounding practice (PGP-R)	●	●	●	■	-Typically, attenuation decreases at high frequency
Poor connection of short circuit cables (PCSC-R)	■	●	●	●	-Deviations at low frequencies with different attenuations
Instrument (MI-R)	●	●	●	●	- No effect
Noise (N-R)	■	■	●	●	-Deviations at power frequency and near 1st resonance point

4 Evaluation of numerical indicators

This chapter collects all the commonly used numerical indices on a single platform and compares and evaluates them based on their characteristics to announce the most appropriate indices as the standard ones for objective interpretation of the FRA signatures. A winding assessment factor is also introduced which overcomes the drawbacks of conventional numerical indices.

4.1 Case studies for evaluation of indices

To evaluate and compare the characteristics of indices, presented in Chapter 2 2.7.1, different case studies are considered. The case studies are summarized in Table 4-1. The first five cases are implemented in an experimental setup [59], [84]. The experimental setup consists of a single-phase transformer with HV and LV windings. The windings correspond to a medium voltage transformer of 1 MVA. The HV winding is a continuous disk winding and the LV winding is a layer winding as shown in Figure 4-1(a)–(c). In the experimental setup, three commonly occurring mechanical faults; axial displacement (AD), radial deformation (RD), and disk space variation (DSV) are implemented in various steps as described in Table 4-1.

In case 6, a cast-resin 160 kVA, 10.5 kV/400 V transformer is employed. The height of the HV winding is 450 mm including epoxy insulation. AD fault is applied by inserting spacers below the HV winding as shown in Figure 4-1(d) [53].

In case 7, a three-phase 3 MVA, 60 kV/6.9 kV transformer is employed. A short circuit current is injected into LV winding by shorting the HV windings of the three phases to investigate the mechanical strength of the aged power transformer. For the U-phase, a short-circuit current of 1.91 kA is injected 3 times based on the IEC60076-5-2006 [85] standard. For the V-phase the current is increased from 1.91 kA with a 10% increment. The short circuit current is injected 9 times and the maximum short-circuit current is 3.44 kA. For the W-phase, the short circuit current of 3.09 kA is injected 3 times. The TFs of the windings of the U-phase and the V-phase have not changed due to the short-circuit tests. The TFs of the W-phase of the LV winding measured after the third short-circuit test have changed.

After dismantling, it was found that the LV winding of the W-phase was deformed [86]. In the eighth and ninth cases, the TFs of a 27 MVA, 154 kV/10.5 kV transformer are measured. Afterward, radial deformation and axial displacement faults are implemented on HV windings, and TFs are recorded for healthy and deformed phases as described in [86].

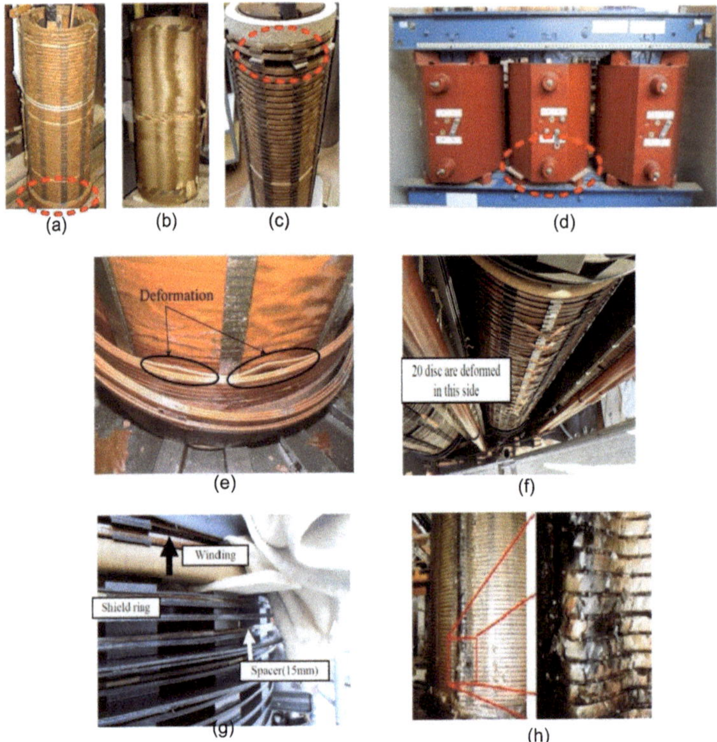

Figure 4-1: Case studies for evaluation of numerical indices

(a) Axial displacement in experimental setup of windings

(b) Radial deformation of LV winding in the experimental setup

(c) Disc space variation in HV winding in the experimental setup

(d) Axial displacement of HV in cast resin transformer, 160kVA, 10.5kV/400V

(e) Deformation of phase-W LV winding after short circuit test, 3 MVA, 60 kV/6.9 kV unit

(f) Axial displacement of HV winding, 27 MVA, 154 kV/10.5 kV unit

(g) Radial deformation of HV winding, 27 MVA, 154 kV/10.5 kV unit

(h) Radial buckling of MV winding, 200 MVA, 220 kV/110 kV transformer

In the last case, TFs of a 200 MVA, 220/110/30 kV transformer are measured in two different time periods [9]. The reference TFs were measured in 1999 and repeated measurements were performed in 2005. After comparison, it was noticed that the TF of MV winding of U-phase has changed. After dismantling, it was found that the MV winding of the U-phase is radially deformed through the full winding height as shown in Figure 4-1(e).

Table 4-1: Case studies for evaluation of numerical indices

Case	Description
1	1 MVA distribution transformer windings 10 steps of axial displacements, each step 5 mm
2	1 MVA distribution transformer windings 5 steps of axial displacement, each step 1 mm
3	1 MVA distribution transformer windings 5 steps of radial deformation, each step 2.5 mm
4	1 MVA distribution transformer windings 5 steps of disc space variation, disc 2-3, each step 1 mm
5	1 MVA distribution transformer windings 5 steps of disc space variation, disc 3-4, each step 1 mm
6	160kVA, 10.5kV/400V cast-resin transformer 5 steps of axial displacement, each step 2 mm
7	3 MVA, 60 kV/6.9 kV transformer Short circuit test
8	27 MVA, 154 kV/10.5 kV transformer Radial deformation of HV winding
9	27 MVA, 154 kV/10.5 kV transformer Axial displacement of HV winding
10	200 MVA, 220/110/30 kV transformer Radial buckling of MV winding

4.2 Criteria for evaluation of indices

Mainly, six criteria are proposed for evaluation of numerical indices, i.e., monotonicity, linearity, sensitivity, data-size dependency, and index ratio. The definition of each criterion and the evaluation of all the indices for different fault steps in all cases are presented in the following sections.

4.2.1 Monotonicity Check

Monotonicity is a property of an index, which characterizes the increasing and decreasing behavior of an index. For a monotonic index, the value of the index always increases or decreases with the increasing difference between the two curves. A good index must exhibit monotonic behavior against the increasing degree of the fault since more severe faults increase the deviations in the TFs and, consequently, an index should possess higher values for larger deviations.

Typical behavior of some monotonic and non-monotonic indices are shown in Figure 4-2. Here, the amounts of four indices against different extents of RD for EE-SC measurement from case 3 are shown. The amounts of indices *SD* and *SSE* are increasing with the increasing degree of the fault and, hence, called monotonic. While indices *SSD* and *CD* have decreasing behavior in two or more change steps, hence, called non-monotonic. From Figure 4-2, it is obvious that non-monotonic indices cannot define the extent of change between two TFs. All indices presented in Chapter 2, are evaluated for this criterion. The first six cases from Table 4-1 are used for monotonicity check, in which indices are evaluated in the frequency range of 10 kHz to 1 MHz and four configurations are considered. A good index must hold monotonic behavior in all cases even in all configurations. All the indices that hold even one non-monotonic behavior are discharged from the list of appropriate indices. Under this criterion, the indices including *CCF, CC, LCC, SD, CSD, ED, RMSE, ALSE, SE, SSE, SDA, MM,* and *JD* are monotonic and these indices will be evaluated for the next criteria. The monotonicity results of all the indices are presented in Appendix C.

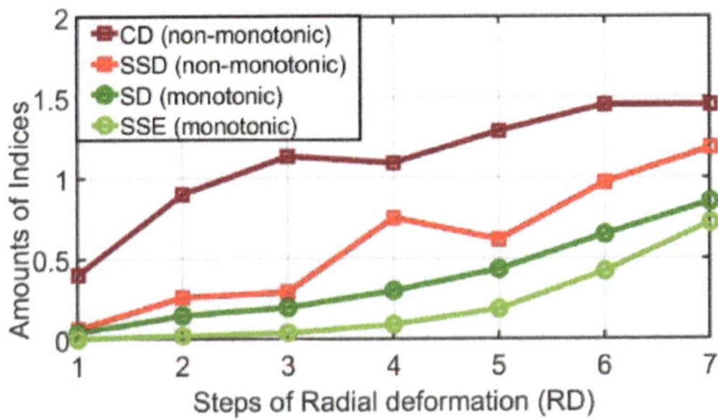

Figure 4-2: Monotonic and non-monotonic behavior of indices against different fault extents

4.2.2 Linearity Check

Linearity is a property of a mathematical relationship, which means that it can be graphically represented as a straight line. To understand the linearity of the numerical indices, the indices are normalized and plotted against different AD fault steps (case 1) in a single graph as shown in Figure 4-3a. It can be seen that different indices have different behavior against each fault step, it is important to mention that the AD fault step is linear. The linearity of the indices is essential for quantitative diagnosis and it is crucial to define limits. To quantify the linearity of the indices, a regression analysis (RA) is performed which is a statistical tool, to define, how much two variables are linearly coupled. The fault steps and the amounts of the indices are taken as input variables. The output of the RA is the coefficient of determination that ranges from 0 to 1, where output equals 1 indicates a complete linear relation. The results of the linearity check are presented in Figure 4-3b. The coefficient of determination reported here is the average of six cases (1–6). The results show that indices *ASLE, LCC, CCF, ED, SD, CSD, RMSE, SDA, MM, SE,* and *JD* exhibit good linear behavior. While indices *CC, SSE, SSD,* and *LSE* are weaker in this respect.

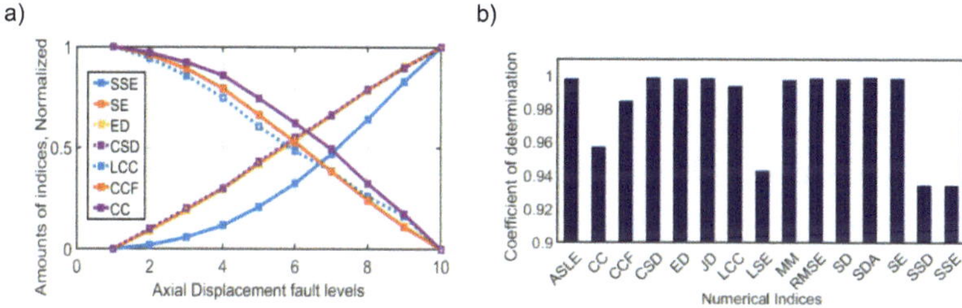

Figure 4-3: Linearity check

(a) Behavior of some indices towards different extents of axial displacement fault

(b) Coefficient of determination as a measure of linearity of numerical indices

4.2.3 Sensitivity Check

According to the principle of TF, both the resonant frequency and amplitude can change with the change in the impedance of the transformer. Therefore, an appropriate index must be sensitive to the frequency and amplitude shifts. Ref [51] and [47] explain that *CC* has some drawbacks. For example, if $Y = cX$, and c is a constant, *CC* equals 1, showing no deviation in the traces although the traces are completely different. Afterward, [47] derives *SSRE* and *SSMMRE* from

SSE. However, they demonstrate weak sensitivities to variations around the trough points. To evaluate the sensitivity of numerical indices against frequency and amplitude shifts, a TF is calculated from a simple RLC series circuit in which horizontal and vertical shifts are applied by varying the values of circuit elements. At first, the resonance frequency is reduced by 1%, 3%, 5%, 10%, and 20% by varying the inductance (L) in the circuit. Secondly, the amplitude of the TF is reduced by 1%, 3%, 5%, 10%, and 20% by varying resistance (R) of the series circuit. The results are demonstrated in Figure 4-4a and Figure 4-4b, respectively. To explain the horizontal and vertical sensitivities of the indices, the amounts of three indices (*CC, CCF,* and *LCC*) are compared against different steps of frequency shifts and amplitude changes as shown in Figure 4-5. It can be seen that each index holds a different behavior to detect the frequency and amplitude changes. Moreover, indices are less prone to detect amplitude changes. For example, the amount of *CCF* is 0.91 for a 20% shift of resonance frequency. While for a 20% change in amplitude, the amount of *CCF* is only 0.99.

To compare and quantify the sensitivities of all the indices, the vertical and horizontal sensitivity of each index is calculated for wide (10–100 kHz) and short (20–50 kHz) frequency bands. The results are shown in Figure 4-6. The results show that the indices *CSD, ED, LCC, SE, SDA, SD,* and *RMSE* possess good sensitivity to detect both frequency shifts and amplitude changes. While indices *CCF, CC, JD, SSE,* and *ASLE* are less prone to amplitude changes. It should be noted that sensitivity also depends on the length of the evaluated frequency band. The sensitivity of all the indices is improved by narrowing the evaluated frequency band except for *ASLE* and *RMSE*.

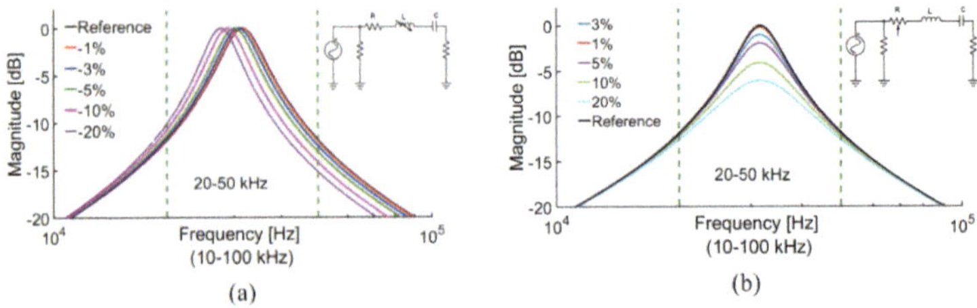

Figure 4-4: TFs of RLC circuit for discussion of horizontal and vertical sensitivities

(a) percentage shift of resonance frequency (Horizontal shift)

(b) percentage change in amplitude (Vertical shift)

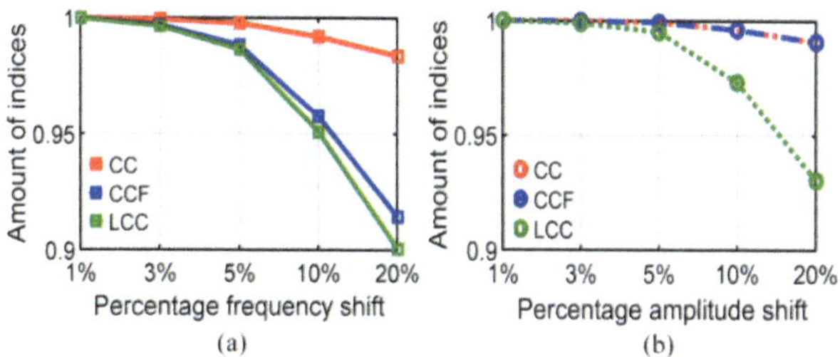

Figure 4-5: Comparison of CC, CCF, and LCC against
(a) frequency shifts
(b) amplitude changes

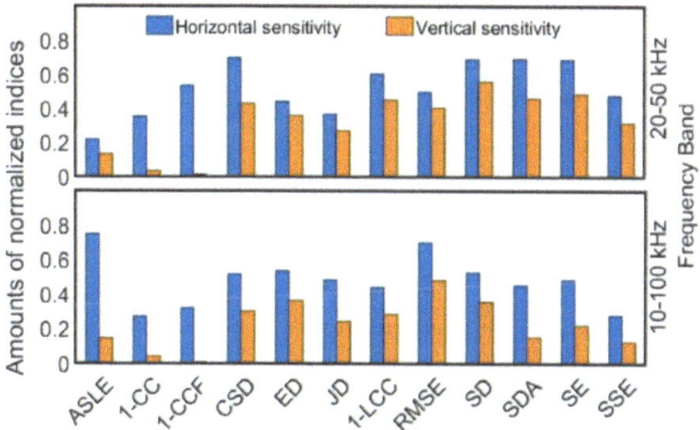

Figure 4-6: Comparison of horizontal and vertical sensitivities of indices for two frequency sub-bands: 10–100 kHz and 20–50 kHz

4.2.4 Data-Size Dependency Check

Different numbers of samples can be measured in the defined frequency range. In the international standard IEC 60076-18 [11], the data size and the measurement frequency resolution are not specified in detail, only the minimum frequency resolution is specified, which is 200 points per decade. In this criterion, the performance of the indices is evaluated for different data sizes. The data size is changed by changing the frequency resolution from 100 to 400 points per decade. For simplicity, the frequency resolution of 100% is chosen to the minimum frequency resolution (200 points per decade) specified in the IEC 60076-18 standard. For this purpose, the FRA results from the cast resin transformer (case 6) are employed, where the response is measured from 10 Hz to 1 MHz. Initially, the TFs are measured for different frequency resolutions (50% to 200%) for the healthy state of the windings. Subsequently, the measurements are repeated when HV winding is displaced up to 8 mm. Table 4-2 shows the percentage change of the indices due to different frequency resolutions. It can be seen that indices *ED, JD, RMSE, SD, CSD,* and *SSE* have a huge impact of the data size. Consequently, these cannot be used to define the extent of a mechanical fault when measurements are performed under different frequency resolutions. While the indices *CC, CCF, LCC, SDA, SE,* and *ASLE* have the least influence on the data size. Hence, these indices can be satisfactorily used to determine the extent of change in TFs when measurements are performed with different frequency resolutions (50% to 200%) and with different data sizes.

Table 4-2: Percentage change of indices due to the change of data size of measurement

Frequency resolution	Data size	ΔCC (%)	ΔCCF (%)	ΔLCC (%)	ΔSD (%)	ΔCSD (%)	ΔED (%)
100 %[(1)]	1000	--	--	--	--	--	--
200 %	2000	-0.007	-0.015	-0.015	7.66	7.65	-30.61
150 %	1500	-0.008	-0.016	-0.016	8.32	8.31	-12.30
50 %	500	0.006	0.011	0.011	-5.91	-5.86	25.14
(1) 100 % Frequency Resolution (200 data points per decade) is taken as a reference							
Frequency resolution	**Data size**	**ΔJD (%)**	**ΔSDA (%)**	**ΔRMSE (%)**	**ΔSE (%)**	**ΔASLE (%)**	**ΔSSE (%)**
100 %[(1)]	1000	--	--	--	--	--	--
200 %	2000	14.70	3.85	7.64	4.83	3.79	14.69
150 %	1500	15.92	3.24	8.30	2.23	1.14	15.91
50 %	500	-12.06	0.90	-5.85	-2.56	-2.7	-12.05

4.2.5 Index Ratio Check

Another important characteristic of the numerical indices is the ability to exhibit high relative change between healthy and faulty states of the transformer. As the basis of FRA analysis is a comparison with reference measurements, and therefore the repeatability of the measurement exactly as reference measurement is an important factor to consider. Consequently, the indices must also be checked against repeated measurements or measurements with healthy phases of the faulty transformers. If an index possesses a high relative change between healthy and faulty cases then it can satisfactorily detect the deviations in TFs due to mechanical faults. To evaluate this condition, the index ratio criterion is proposed which quantifies the relative change of the indices between a healthy (which is a repeated measurement) and a faulty phase/winding. An appropriate index must possess a high relative change between healthy and faulty phase/winding. The index ratio is calculated from the given relation:

$$Index\ Ratio = \frac{Value\ of\ index\ in\ faulty\ phase}{Value\ of\ index\ in\ healthy\ phase} \qquad (4\text{-}1)$$

The results of the index ratio are presented in Figure 4-7. The index ratio of five cases (6–10) is reported here and evaluated for two frequency sub-bands including 10 kHz–500 kHz and 500 kHz–1 MHz. It can be seen from the results that *LCC* and *CCF* exhibit high relative change. However, looking back at Figure 4-6, *CCF* is least sensitive to amplitude changes.

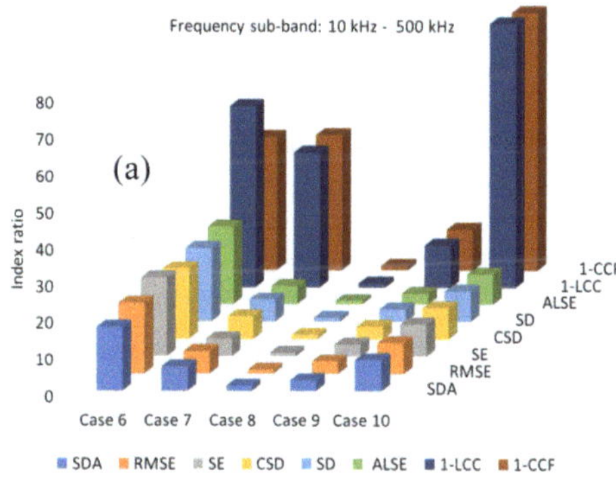

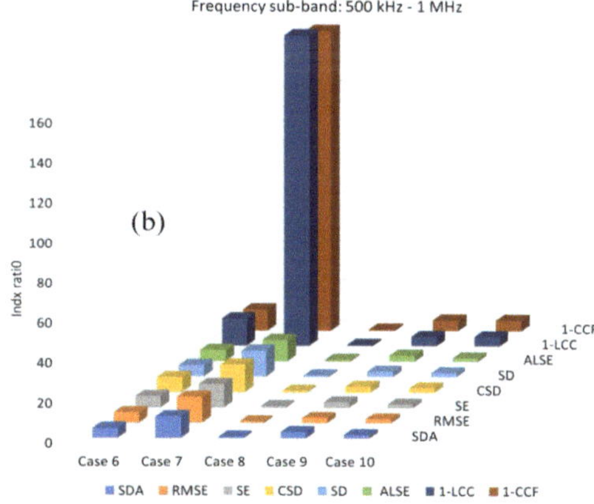

Figure 4-7: Comparison of index ratio of indices for two frequency bands
(a) 10–500 kHz
(b) 500 kHz–1 MHz

4.2.6 Summary on evaluation of indices

Table 4-3 summarizes the results of all the criteria. It can be seen that different indices hold different behavior within each criterion. The monotonicity check result shows that most of the indices are monotonic. For linearity check, almost all monotonic indices show good linearity. Only four indices are weaker in this aspect. In the case of sensitivity check, the indices including *CSD, ED, LCC, SD, SDA, RMSE,* and *SE* stand out from others in terms of showing good sensitivity to detect both frequency shift and amplitude changes. For the data size dependency, the indices *CC, CCF, LCC, SDA, SE,* and *ASLE* have the least influence of different numbers of data points. In index ratio criteria, the indices *LCC* and *CCF* show high relative change between healthy and faulty phase/winding.

In summary, based on the data presented here and the results of all the criteria, this study recommends the four indices *LCC, SDA, SE,* and *CCF* as the most appropriate ones that can satisfactorily define the extent of change between two TFs. While *CSD* and *SD* also show reasonable performance but they exhibit slightly higher data-size dependency and lower index ratio. Along with these criteria, there are also some other points that must be unified regarding the indices to be used as a standard method. First, the definition of a standardized method is required for indicating frequency bands in different transformers to unify the implementation of indices. Secondly, the frequency resolution and the data size in the measurement should be fixed. However, this issue can be resolved by using only the recommended indices that have the least influence of data size. Thirdly, the definition of the threshold values must be established.

Table 4-3: Summary of the criteria for selection of appropriate numerical indices

		Numerical Indices											
		CCF	CC	LCC	JD	ED	SD	CSD	SE	SSE	RMSE	ALSE	SDA
Criterion	Monotonicity	✓	✓	✓	✓	✓	✓	✓	✓	✓	✓	✓	✓
	Linearity	● green	● red	● green	● green	● green	● green	● green	● green	● red	● green	● green	● green
	Sensitivity	◆ red	◆ red	◆ green	◆ red	◆ green	◆ green	◆ green	◆ green	◆ red	◆ green	◆ red	◆ green
	Data-size dependency	■ green	■ green	■ green	■ red	■ red	■ red	■ red	■ green	■ red	■ red	■ green	■ green
	Index Ratio	▲ green	-	▲ green	-	-	▲ red	▲ red	▲ red	-	▲ red	▲ red	▲ red

Nomenclature		
■ (shaded): Recommended Indices	● green: Linear; ● red: Less linear	✓: Monotonic
▲ green: High; ▲ red: Low	◆ green: Sensitive; ◆ red: Less sensitive	

4.3 Transformer Winding Assessment Factor

Focusing on the challenges of fixed frequency sub-bands for the application of numerical indices, in the framework of this research, a winding assessment factor is proposed which is a vector-based method using a sliding window approach [79], [87]. In this method, the whole frequency response is scanned in order to cope with the problem of defining fixed frequency sub-bands. The proposed method is based on the fact that the degree of deviation is quantified by evaluating the mean deviation of differences between two variables and this deviation is then standardized with the number of samples, hence, called Standard Deviation of Difference *(SDD)*. In this method, *SDD* is calculated in a window with a specific window size. This window is moved from the starting frequency to the ending frequency of the TF with a specific window step (Wstep=1). In this way, the entire frequency spectrum of TF is scanned and *SDD* is calculated in each window. Consequently, a vector is obtained characterizing the differences between two TFs as a function of frequency. Hence, the calculated *SDD* and measured transfer functions can be presented on the same graph. The winding assessment factor is calculated as:

$$\overrightarrow{SDD(i)} = -2\left(\sqrt{\frac{\sum_{j=1}^{WS}\left(Z(j) - \overline{Zw(i)}\right)^2}{WS-1}}\right)$$

$$\overrightarrow{Z(i)} = X(i) - Y(i)\,, \quad \overline{Zw(i)} = \overline{Xw(i)} - \overline{Yw(i)}$$

$$\overline{Xw(i)} = \frac{1}{WS}\sum_{j=1}^{WS} X(j)\,, \quad \overline{Yw(i)} = \frac{1}{WS}\sum_{j=1}^{WS} Y(j) \qquad (4\text{-}2)$$

$$i = 1, 2, 3 \ldots N$$

$$WS = 10 + 6\left(\frac{f_{res} - 200}{200}\right)$$

Where, X and Y are the magnitude vectors of reference and measured TFs, respectively. $X(i)$ and $Y(i)$ are the ith elements of these vectors. $\overline{Xw(i)}$ and $\overline{Yw(i)}$ are the means of the ith window. f_{res} is the number of data points per decade and WS is the window size. In the international standard IEC 60076-18 [11], the minimum measurement frequency resolution is specified, which is 200 points per decade. To consider the effect of different frequency resolutions, the window size is made variable and is calculated from Equation (4-2). In this way, the effect of different frequency resolutions is also considered. Figure 4-8 illustrates the basic principle of the presented method. The representation of the proposed method for identification of the degree of deviation between two TFs of a three-phase transformer is shown in Figure 4-9. It can be seen that the proposed method has the following advantages:

- Increased fault detectability
- Removes the issue of fixed frequency sub-bands for FRA interpretation
- Plotting deviation between two TFs as a function of frequency
- Possibility to set a threshold to the lowest value of *SDD* function, i.e., *MSDD*
- Possibility of fault classification

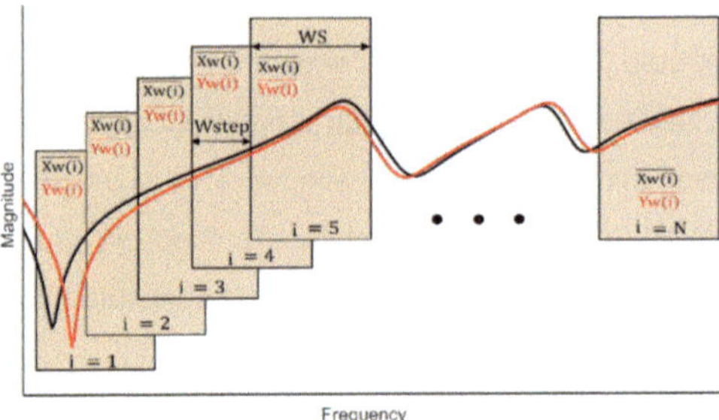

Figure 4-8: Basic principle of winding assessment algorithm

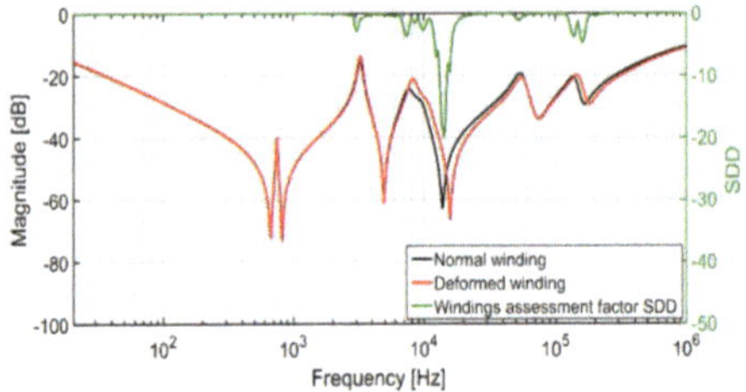

Figure 4-9: Representation of the winding assessment factor SDD for a three-phase transformer

4.4 Diagnostic Criterion of Transformer Winding

To demonstrate the performance of the winding assessment factor, 11 case studies from 8 different power transformers are selected from database case studies which are also presented in chapter 3. In order to evaluate the sensitivity of the method against different faults, only the cases provided with reference measurements are considered here. While the cases provided with phase-to-phase measurements are therefore avoided. A conceivable way to assess the performance of this method is to look at the value of the factor in healthy and deformed cases. For this purpose, TFs from the healthy phases and the cases that provided successful short circuit tests are also included. Table 4-4 illustrates the details of each case study for damaged and healthy transformers.

The TFs of the selected cases (Table 4-4) along with the calculated winding assessment factor *(SDD)* are shown in Figure 4-10 - Figure 4-17. For fault assessment, the minimum value of the *SDD (MSDD)* should be focused which is also displayed in Figure 4-10 - Figure 4-17. It can be seen that the winding assessment factor *(MSDD)* can identify the degree of deviation between FRA traces with increased sensitivity. However, some regions should be disregarded during interpretation. As in open-circuit measurements at the low-frequency region, the deviation between frequency responses is due to different remanence conditions of the core. These regions are indicated by the arrow and symbol "**##**".

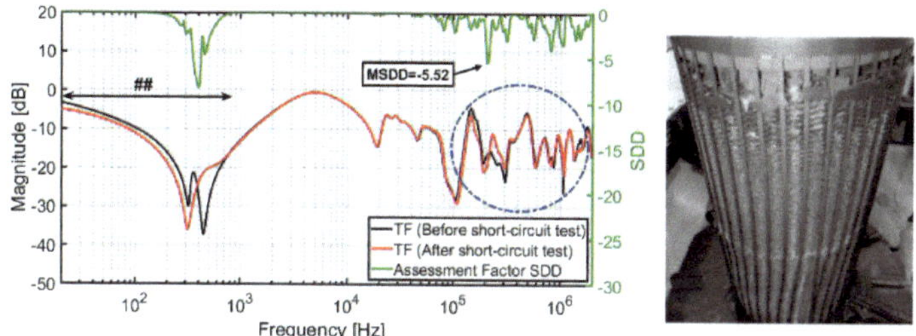

Figure 4-10: Application of assessment factor with EE-OC TFs of tertiary winding before and after the short circuit event (left) and fault inspection (right) – case 1

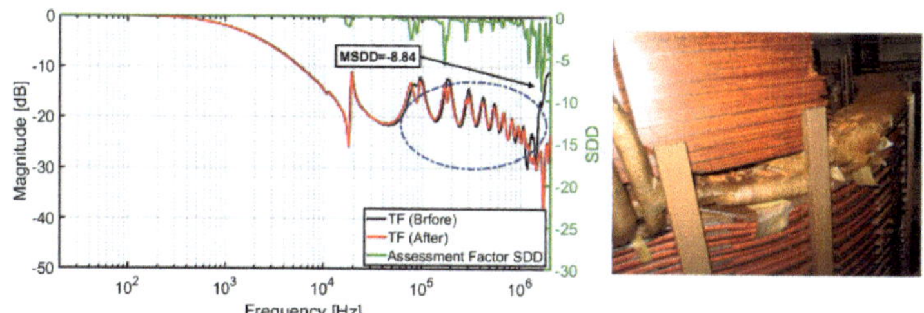

Figure 4-11: Application of assessment factor with EE-SC TFs of LV winding (left) and fault inspection (right) – case 2

Table 4-4: Selected cases for the application of winding assessment factor

Case	Transformer details		Description
	Rating	Detail	
1	450 MVA, 315/120/12.5 kV	3-phase, auto-transformer	Lead movement, axial collapse, and spiraling
2	21.6 MVA, 166/10 kV	Single-phase	Axial movement of the LV winding
3	250 MVA, 400/155 kV	3-phase, auto-transformer	Twisting of the phase C tap winding
4	27 MVA, 154/10.5 kV	Y-Δ	Axial displacement of HV winding, phase U
5	27 MVA, 154/10.5 kV	Y-Δ	Radial deformation of HV winding, phase V
6	1 MVA	Single-phase	Buckling of LV winding
7	3 MVA, 60/6.9 kV	Δ- Δ	Buckling of LV winding after short circuit test
8	3 MVA, 60/6.9 kV	Δ- Δ	Successful short circuit test
9	250 MVA, 400/155 kV	3-phase, auto-transformer	Successful short circuit test
10	47 MVA, 120/26.4 kV	Y- Δ	Successful short circuit test
11	120 MVA, 220/63/10.5 kV	Y- Y- Δ	Successful short circuit test

All the cases presented in Table 4-4 (healthy and deformed) are also tested with existing standards and practices, i.e., R_{xy}, *E,* and *CCF*. Table 4-5 summarizes the calculated values of R_{xy}, *E, CCF,* and *MSDD* for healthy and deformed cases. It is noteworthy that frequency sub-bands for the calculation of R_{xy}, *E*, and *CCF* are the same as mentioned in [45], [61], and [46], respectively. From Table 4-5, it is evident that out of 7 deformed cases, R_{xy} could only detect deformation in one case (case 2). Whereas, in cases 1 and 3, R_{xy} has detected deformation in the low-frequency region. While referring to Figure 4-10 and Figure 4-12, it is evident that the deviation between TFs is pronounced in the high-frequency region. It means R_{xy} misinterpreted these cases due to the core saturation effect. Consequently, R_{xy} fails to show any deformation in cases 4, 5, 6, and 7, indicating a 'normal- winding'. Moreover, out of 4 normal cases, R_{xy} misinterpreted 3 cases (cases 9, 10, and 11), indicating 'deformed transformer'. Likewise, *E* could not detect any faulted case, indicating 'normal-transformer' for cases 1-7. Similarly, *CCF* could only detect obvious faults in two cases (cases 1 and 2) and marginal deformation in case 7. While it also fails to show deformations in cases 4, 5, and 6, indicating a 'normal winding'. There can be two possible reasons for the impoverished performance of these criteria in some cases:

- The definition of the general fixed frequency sub-bands is a gross estimation, valid in some cases but not in others
- Low sensitivity of the indicator employed in these criteria

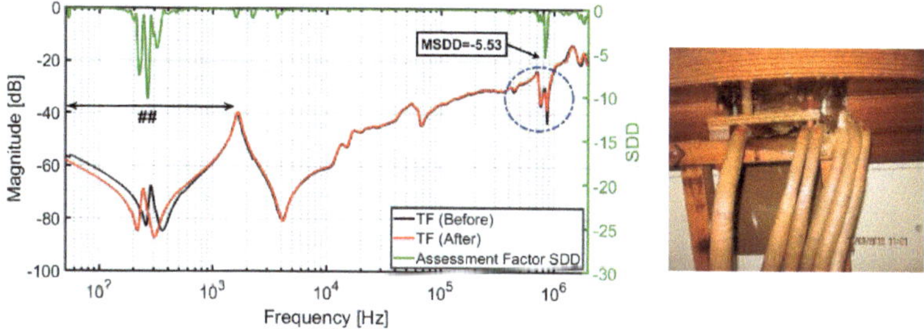

Figure 4-12: Application of assessment factor with EE-OC TFs of HV winding before and after short circuit event (left) and fault inspection (right) – case 3

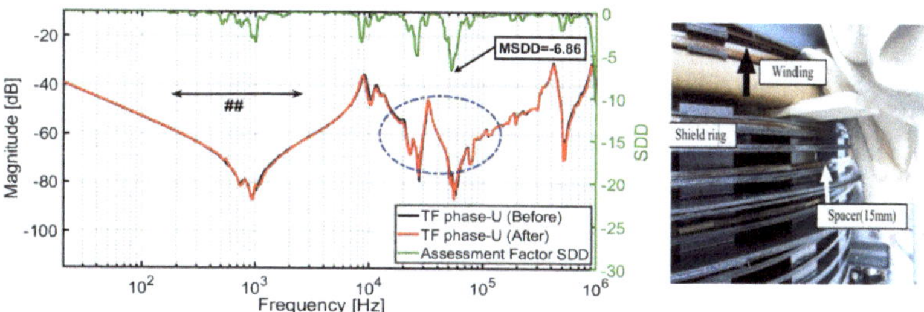

Figure 4-13: Application of assessment factor with EE-OC TFs of HV winding before and after axial displacement (left) and fault simulation (right) – case 4

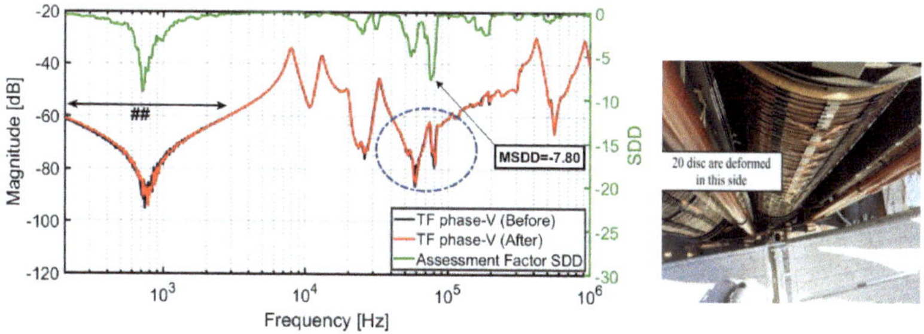

Figure 4-14: Application of assessment factor with EE-OC TFs of HV winding before and after fault (left) and fault simulation (right) – case 5

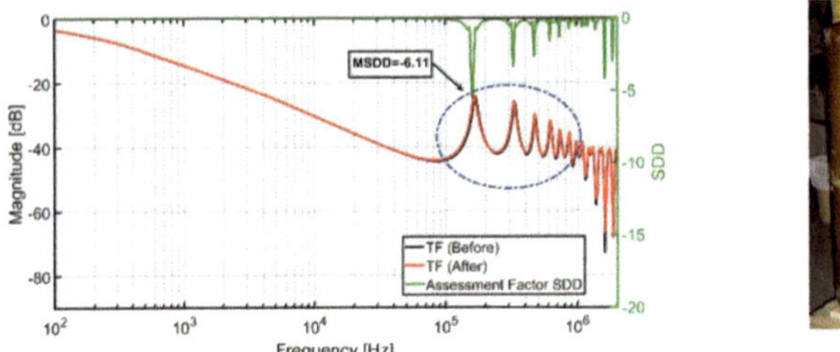

Figure 4-15: Application of assessment factor with EE-SC TFs of HV winding before and after fault (left) and fault simulation (right) – case 6

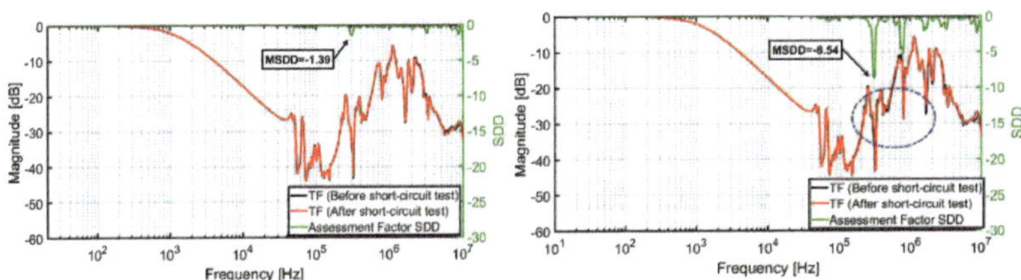

Figure 4-16: Assessment factor application with SC TFs of LV winding before and after successful short-circuit test - case 8 (left) and unsuccessful short circuit test - case 7 (right)

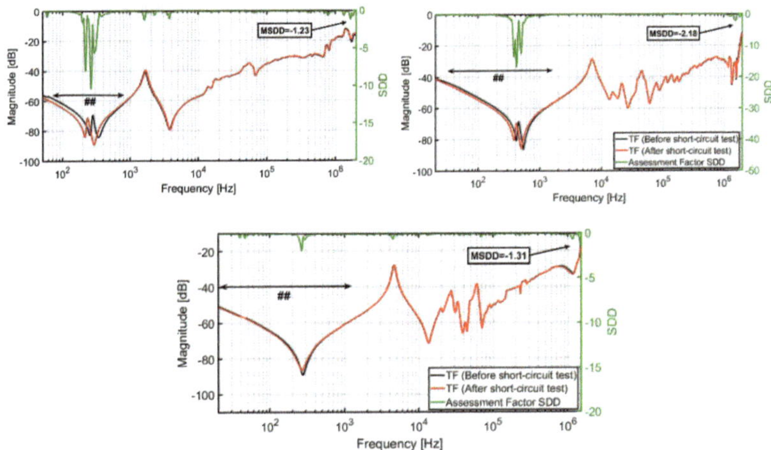

Figure 4-17: Application of assessment factor to the healthy transformers

(a) Successful short circuit test – case 9

(b) Successful short circuit test – case 10

(c) Successful short circuit test – case 11

Table 4-5: Comparison of winding assessment factor with existing international practices for selected case studies

No.	Transformer condition	Description of case studies	Existing standards and international practices								SDD
			R_{xy}			E	CCF				
			LF	MF	HF		LF1	LF2	MF	HF	
1	Deformed	Lead movement, axial collapse, and spiraling	1.18	1.05	1.01	1.67	0.88	0.99	0.95	0.96	-5.52
2	Deformed	Axial movement of the LV winding	2.41	0.95	1.12	1.14	1.00	0.99	0.89	0.96	-8.84
3	Deformed	Twisting of the phase C tap winding	1.53	2.72	1.61	0.34	0.91	0.99	0.99	0.98	-5.53
4	Deformed	Axial displacement of HV winding, phase U	2.50	2.54	1.68	1.13	0.99	0.99	0.99	0.99	-6.86
5	Deformed	Radial deformation of HV winding, phase V	2.29	2.76	3.52	0.97	0.98	0.99	0.99	0.99	-7.80
6	Deformed	Buckling of LV winding	4.29	1.58	1.46	0.96	1.00	1.00	0.98	0.97	-6.11
7	Deformed	Buckling of LV winding	4.89	1.87	1.19	0.86	1.00	1.00	0.99	0.96	-8.54
8	Normal	Successful short circuit test	5.19	3.50	3.07	0.15	1.00	1.00	0.99	0.99	-1.39
9	Normal	Successful short circuit test	1.59	2.60	2.85	0.19	0.92	0.99	0.99	0.99	-1.23
10	Normal	Successful short circuit test	1.75	3.58	1.85	0.15	0.96	0.99	0.99	0.99	-2.18
11	Normal	Successful short circuit test	2.82	4.02	0.55	0.12	0.99	0.99	0.99	0.96	-1.31

Normal	Deformed	Marginal

In contrast, Table 4-5 also shows the minimum values of the *SDD* indicator for each case. It is evident that the *SDD* indicator has detected all the faulted cases (1-7) with good sensitivity. Thus, the proposed method has the ability to detect faults with high sensitivity, and it also resolves the problem of fixed frequency sub-bands. It is worth noting that *MSDD* has lower values for the cases with deformations than for those which are normal as shown in Figure 4-18. Thus, it is possible to set a threshold to the minimum value of *SDD* as a diagnostic criterion, disregarding the frequency ranges. For example, *MSDD=-5* can be employed as a threshold for time-based FRA comparisons. *MSDD<-5*, indicates abnormal winding whereas, *MSDD>-5* indicates normal winding as shown in Figure 4-19. It should be noted that in all the presented case studies, time-based comparisons are performed. A similar analysis can be performed for type and construction-based comparisons and permissible threshold values can be calculated using the proposed *SDD* indicator.

The presented method also gives the possibility of classifying different types of faults. For example, it is possible with the application of assessment factor to identify the most affected frequency ranges under different faults as shown in Figure 4-20. This analysis would ease the objective interpretation of transformer FRA results. However, it is important to note that the frequency ranges mentioned here are gross estimations and may vary with the size and rating of the windings.

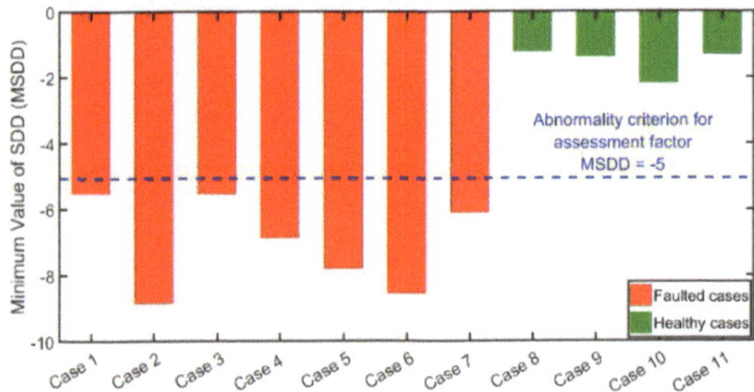

Figure 4-18: Minimum values of SDD (MSDD) for selected case studies; Case 1-11

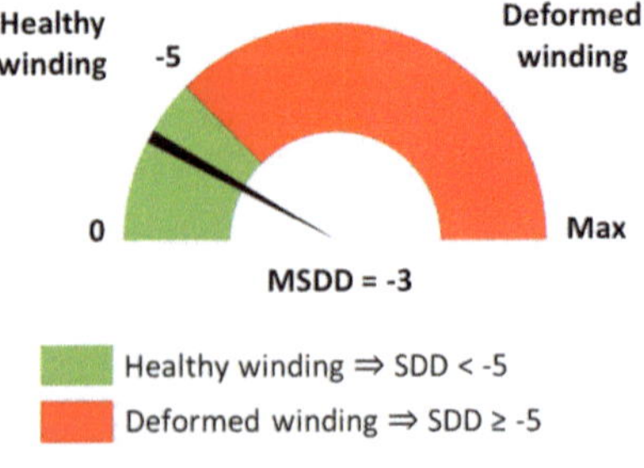

Figure 4-19: Transformer winding abnormality criterion

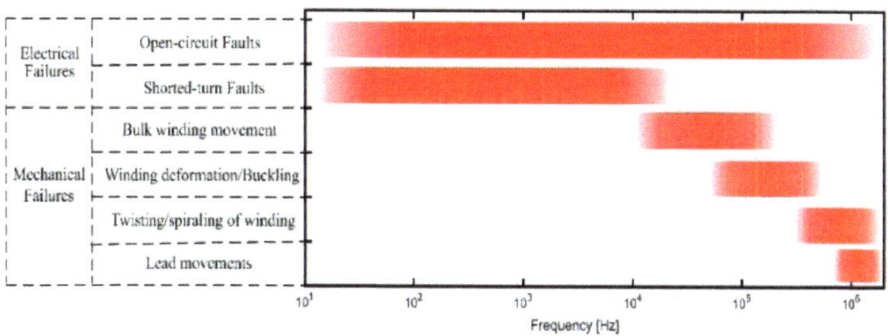

Figure 4-20: Affected frequency ranges under different electrical and mechanical faults

4.5 Summary

In summary, based on the results of all the criteria, four indices, i.e., *LCC, SDA, SE,* and *CCF* show appropriate performance in detecting the changes between two TFs. However, it is difficult to establish the threshold values for these indices as a diagnosis criterion. As these indices are evaluated in different frequency sub-bands, while frequency sub-bands cannot be generalized. Here comes the significance of the proposed winding assessment factor. It determines the degree of deviation by evaluating the mean deviation of the difference between the two transfer functions. The presented method overcomes the drawbacks of conventional numerical indicators and gives the following advantages for the assessment of FRA results.

- Enhanced fault detectability
- Resolves the problem of frequency sub-band division
- Plot deviations on the same graph as a function of frequency
- Possibility to set a threshold, i.e., *MSDD*
- Possibility for fault classification

It is important to mention, that the method is tested on 11 cases with 8 transformers, more case studies from the field are necessary to verify its performance under different fault conditions. Especially, it is necessary to investigate the method with the data from repeated measurements at healthy transformers.

5 Transformer High-frequency model

This chapter presents 3D High-frequency modeling of a 3-phase transformer and validation with the experimental setup. Simulation of different electrical and mechanical faults is also included in this chapter.

There exists a wide range of models (black box, white box, grey box) and procedures for parameter estimation as mentioned in Chapter 2. However, many of those procedures have been validated by analyzing steady-state conditions only and some used lumped parameter models with inadequate parameter representation that cannot be extrapolated to transformers of different winding types, ratings, or when the phenomena include high frequencies. These concentrated parameters models are limited to a certain frequency range due to the difficulties in calculating the parameters needed to build a turn-to-turn model.

The Finite Element Method (FEM)/Finite Integration Technique (FIT) compared to the analytical method is more flexible and has many applications in the simulation of transformer behavior. By using FEM/FIT and considering the structural details of a transformer, it is possible to calculate the frequency-dependent parameters of windings in detail, and simulation of precise and accurate faults is also conceivable. Looking at the current dilemma in the FRA field, there are limited case studies available to understand the effect of different faults, and due to the destructive nature, it is not possible to apply the real mechanical deformations in the transformer windings to obtain the data. In this regard, transformer FRA simulation models can be useful to investigate the effect of different failure modes in different transformers.

5.1 High-Frequency Transformer Modelling

The main idea of using a 3D high-frequency model is to emulate the real transformer and FRA measurement operations. CST Microwave Studio is used to model transformer behavior and FRA measurement operations, which is based on FIT. FIT was first proposed by Weiland in 1977 [88]. FIT is a spatial discretization approach to numerically solve electromagnetic fields in the time and frequency domain, which lead to a single solution. FIT is based on the integral form of Maxwell equations. In this method, a computational domain is spatially discretized by a doublet of two computational grids: primary grid G and dual grid $\tilde{G}$ as shown in Figure 5-1. The electric voltages (e) and magnetic fluxes (b) are allocated on the primary grid while the magnetic voltages (h) and dielectric fluxes (d) are allocated on the dual grid. Finally, the spatial discretization of the continuous Maxwell equations is performed on this system of orthogonal computational grids. The discrete equivalent of Maxwell's equations, the so-called Maxwell grid equations are expressed as:

$$
\begin{aligned}
&\mathbf{C}\hat{\mathbf{e}} = -\frac{d}{dt}\hat{\hat{\mathbf{b}}} \\
&\tilde{\mathbf{C}}\hat{\mathbf{h}} = \frac{d}{dt}\hat{\hat{\mathbf{d}}} + \hat{\mathbf{j}} \\
&\mathbf{S}\hat{\hat{\mathbf{b}}} = \mathbf{0} \\
&\tilde{\mathbf{S}}\hat{\hat{\mathbf{d}}} = \mathbf{q}
\end{aligned}
\tag{5-1}
$$

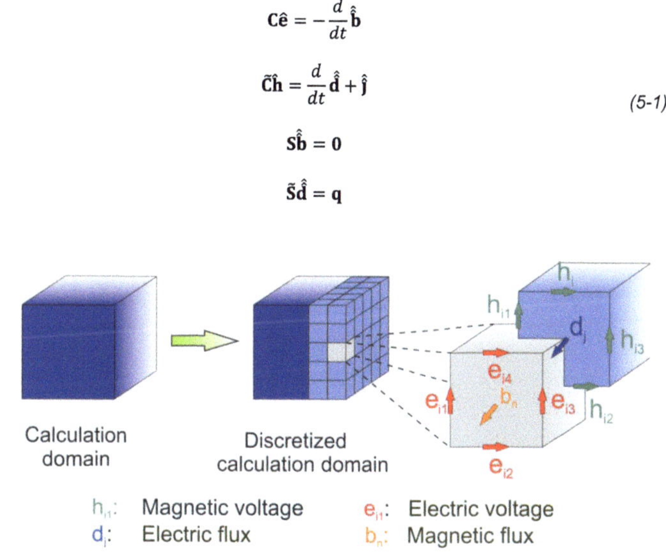

Figure 5-1: Dual discretization grids with allocation of voltage and flux components in the mesh

In these equations, $\mathbf{C}$, $\tilde{\mathbf{C}}$, $\mathbf{S}$, and $\tilde{\mathbf{S}}$ are the topological matrices representing the discrete equivalents of curl and div operators, the tilde matrices belong to the dual grid. In this research, a turn-based 3D high-frequency model of a 3-phase 3 MVA transformer is developed, which consists of HV and LV windings including the insulation papers as well as core and surrounding tank, as shown in Figure 5-2 (tank: hidden). The HV winding is a continuous disk winding with 660 turns in 60 disks and the LV winding is a helical winding with 24 turns and 12 parallel conductors in each turn. The 3D electromagnetic (EM) simulations are performed in a high-frequency solver of CST MW STUDIO [89]. It solves the discrete Maxwell equations in the frequency domain and calculates the electric and magnetic field distributions at different frequencies, hence, it considers the frequency-dependent effects in the core and volume of the conductors. In order to simulate the behavior of the ferromagnetic core, frequency-dependent anisotropic complex permeability is considered. The real and imaginary parts of the complex permeability represent the ability of the core material to conduct the magnetic flux and the losses generated in the core due to eddy currents, respectively. As turn-based geometries of the windings are considered, thus, the HF model provides an accurate determination of turn-based parameters, i.e., self-inductance of each turn, mutual inductance between coils, inter-turn, and inter-disc capacitance, etc. The frequency-dependent losses such as eddy current effects (skin and proximity effect) in the coils and dielectric losses in the insulation structure are also considered.

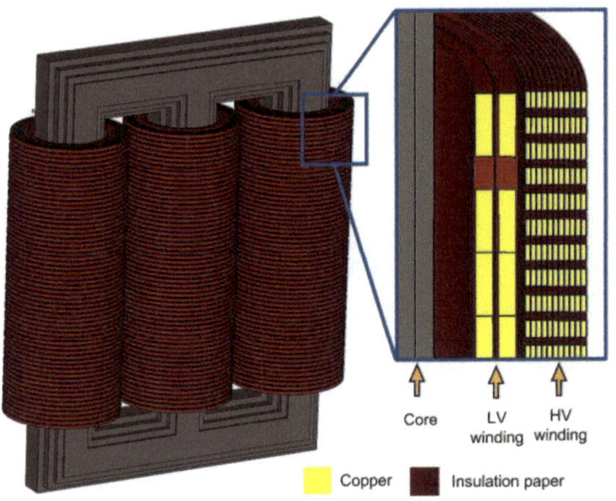

Figure 5-2: Active part geometry of a three-phase transformer in CST

For each 3D EM simulation setup inside CST MW STUDIO two fundamentally different modules exist, i.e., the design module and the schematic module. In CST, FRA simulation is a three-step process. Figure 5-3 explains the flowchart of the algorithm for FRA simulation. Firstly, a 3D turn-based geometric model is created in the design module. Secondly, the broadband frequency-domain electromagnetic field computations are performed in the high-frequency solver where the excitation ports are set at both ends of the windings. The model is discretized by applying tetrahedral mesh. The output of this stage is a multiport network, which characterizes the electromagnetic properties of the 3D geometry model. For a three-phase transformer, a twelve-port network model is obtained. Lastly, the multi-port network is applied to the schematic module, where a sinusoidal voltage source is applied analogously to the real FRA measurements to calculate the FRA traces for different connection schemes. In this way, the FRA traces are directly obtained from the 3D high-frequency model of windings without estimating and solving the lumped parameter circuit models.

Figure 5-4 displays the magnetic field strength at different frequencies, i.e., 20 Hz, 500 kHz, and 1 MHz. As can be seen that magnetic field strength decreases with the increase of frequency. At low frequency, flux penetration in the conductor (winding and core) is high. However, as frequency increases, eddy currents are induced in the conductor (winding and core). Consequently, the flux will be confined to the thin layer near the surface, which reduces the magnetic field strength. It is noteworthy that at 500 kHz and 1 MHz, the flux penetration into the core reduces significantly. Thus, the high-frequency components do not contribute appreciably to the flux in the transformer core as mentioned in [90]. The frequency response of high voltage winding of the HF model for open circuit configuration in a healthy state of the windings is shown in Figure 5-5.

The presented HF model gives manifold advantages in calculating the frequency response of the transformer. First, the HF model considers the frequency-dependency of the winding parameters. Second, transfer functions are calculated directly from the 3D winding model without conversion and solving lumped parameter circuit models. Third, various mechanical faults that are difficult to implement in circuit models, can also be simulated in the presented HF model. Lastly, the frequency response of different types of windings, and various vector groups can be analyzed. Besides, it is also possible to study the effects of windings' different electrical properties on the frequency response of the transformer. For instance, it is possible to change the permittivity and dissipation factor of the oil-paper to simulate the different moisture levels. The proposed modeling method can be useful to study moisture-content identification in different transformer windings with different constructions. Additionally, the proposed model is also useful to study the transient analysis of power transformers, and the calculation of winding overvoltages under different fault conditions.

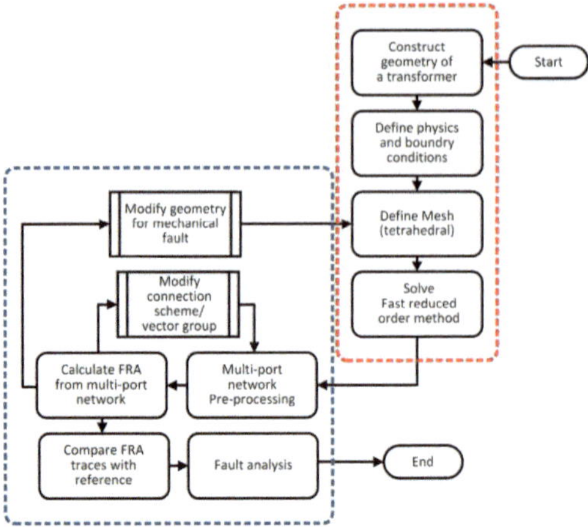

Figure 5-3: FRA simulation algorithm in CST, design module (red), schematic module (blue)

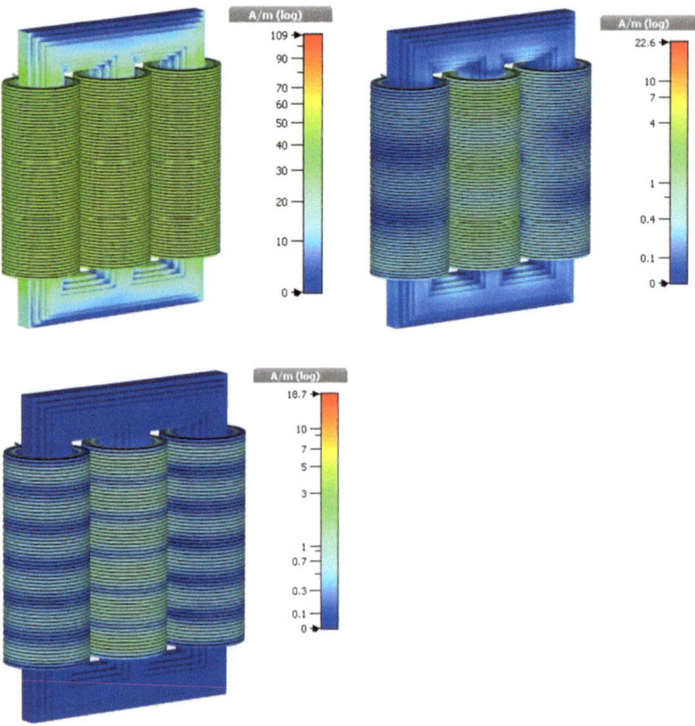

Figure 5-4: Magnetic field distribution in HF model of three-phase transformer at different frequencies: (a) 20 Hz; (b) 500 kHz; (c) 1 MHz

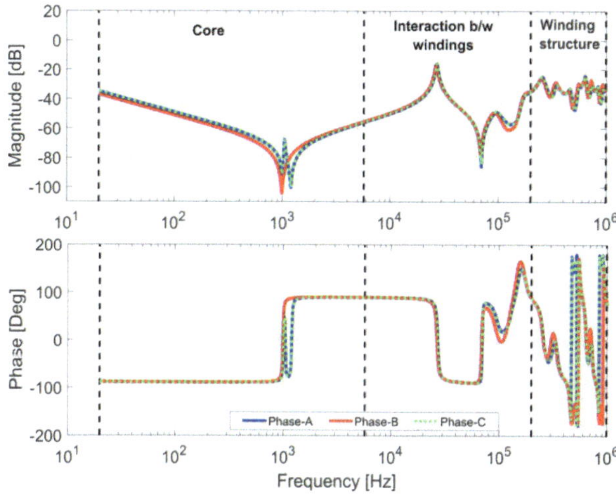

Figure 5-5: HV EE-OC frequency response of HF model of the three-phase transformer model

5.2 Validation of 3D HF Transformer Model

The method is validated with a simple experimental setup, which consists of HV and LV windings. The windings correspond to a medium voltage transformer of 1 MVA. The HV and the LV windings are similar to the three-phase HF model of the transformer, as described in section 5.1. In this setup, two hollow copper cylinders are employed outside and inside of the windings to model the tank and the core, respectively. The experimental setup and the corresponding CST MW Studio model are shown in Figure 5-6. The model was validated with measurements for both healthy and deformed states of windings as shown in Figure 5-7. The performance of this method against different mechanical faults was also proven in [91], [92], [93], [94]. The results of axial displacement (AD) fault in the IIW connection scheme are presented here. Five steps of AD (each step=10 mm) are implemented in the HV winding, in both experimental setup and CST model. The simulation results are compared with measurements as shown in Figure 5-8. The results show good principle agreement of the simulations with the measurements, which proves the applicability of the HF model for the interpretation of transformer frequency response. It is important to mention that it is hard to generate the exact FRA fingerprints of transformers because it demands high accuracy and large computation space. However, the presented model can be used for FRA interpretation studies.

It is important to mention that different types of windings (layer, continuous disc, interleaved disc, etc.) can also be simulated to investigate the effect of winding structure on frequency response. A study on the effect of the type of winding on FRA is presented in Appendix E, where three main types of windings are simulated and their effect on FRA is discussed objectively.

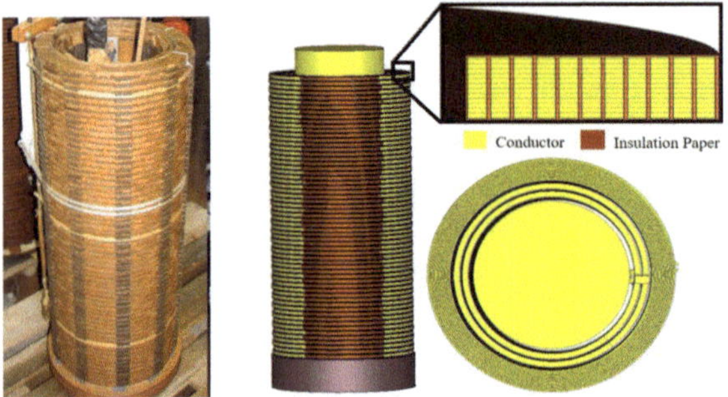

Figure 5-6: Experimental setup (left) and CST model (right)

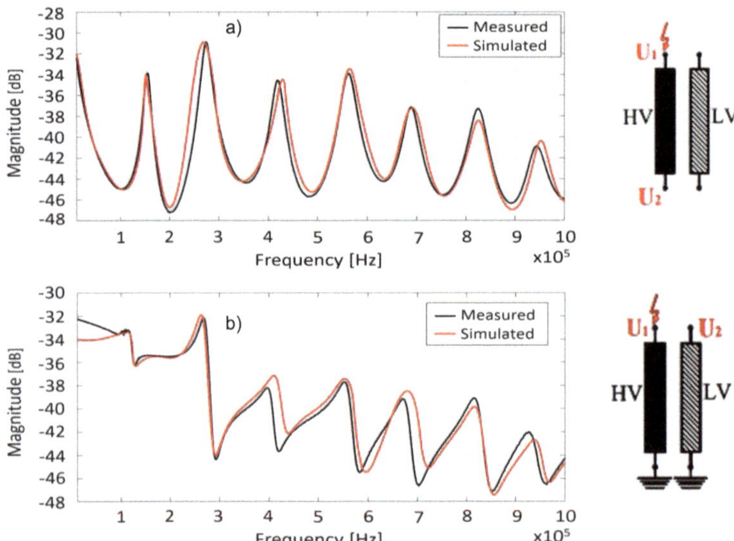

Figure 5-7: Comparison of measured and simulated TFs for the healthy state of windings; (a) EE-OC (b) IIW

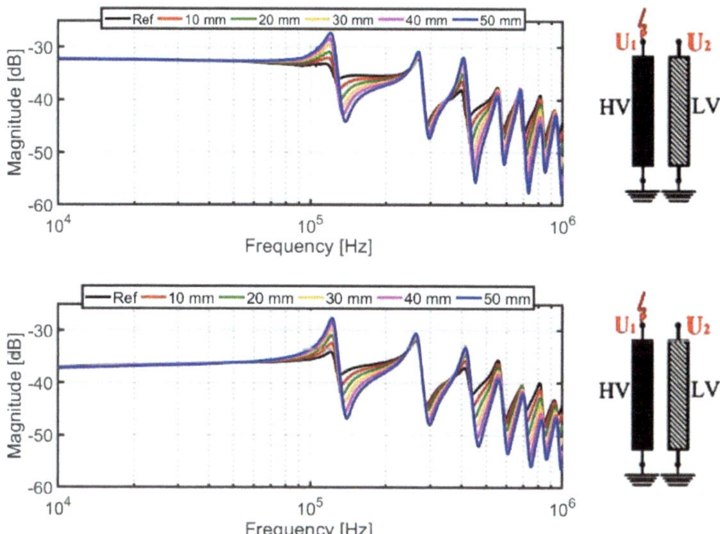

Figure 5-8: IIW TFs for different levels of axial displacement fault; measurement (top) simulation (bottom)

5.3 Fault simulation

In this section, a comprehensive analysis of the simulation of winding electrical and mechanical faults is presented and their effect on FRA test configurations is discussed objectively.

5.3.1 Mechanical Failure Modes

5.3.1.1 Axial displacement fault (Al-M)

Any mechanical failure of the clamping system would allow windings to move in opposite vertical directions relative to one another, it is also referred to as bulk winding movement or telescoping [5]. A schematic representation of the Al-M, in a typical transformer winding, is shown in Figure 5-9. To simulate the Al-M fault, the HV winding of B-Phase is displaced downward by $\Delta h = 30$ mm (~3.5% of its height) as shown in Figure 5-9. Due to the courtesy of the multiport network, it is possible to connect the 3-phases of the model in different vector groups. In this study, the TFs of the YNyn0 vector group are simulated. To discuss the sensitivity of different connection schemes, four connection schemes, i.e., EE-OC, EE-SC, CIW, and IIW are implemented. Figure 5-10 shows the effect of AI-M on TFs of EE-OC and EE-SC connection schemes. While the results of interwinding connection schemes are presented in Appendix F. Under AI-M, the interaction of the windings and hence the mutual couplings (C_w and M_u) are changed. The impact of AI-M is different in different connection schemes and is briefly explained in the following:

HV EE-OC: TF remains unchanged in LFB1. In LFB2, the first anti-resonance point is slightly shifted to the right due to the decrease of the C_w. The effect of ADF is most obvious in the MFB and HFB, where the resonance frequencies are shifted to the right due to the decrease of M_u and C_w.

HV EE-SC: The effect of AI-M can be perceived in the inductive roll-off region. In this region due to the change of impedance, the difference in TFs of healthy and affected winding is 0.7 dB. In MFB and HFB, the effect is similar to HV EE-OC.

IIW (Figure F.1): TFs in LFB1 and LFB2 give a measure of the turn ratio, and it remained unaffected. While TFs in MFB and HFB are significantly changed, new resonance peaks and valleys are observed.

CIW (Figure F.1): The low frequency (LFB1 and LFB2) response of this configuration shows a linear behavior that is dominated by C_w. Due to the decrease of C_w, the TFs are shifted downward with a change of 0.6 dB. Also, the resonance frequencies are shifted to the right due to the decrease of C_w and M_uu. The deviation between TFs is maximum in HFB.

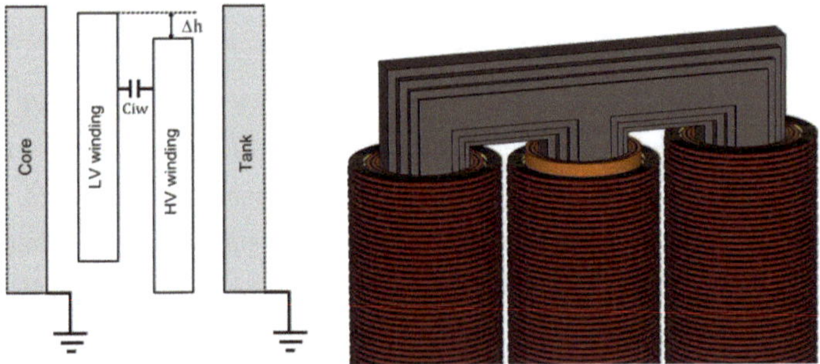

Figure 5-9: Representation of AI-M; schematic (left); B-phase HV winding of 3D HF model (right)

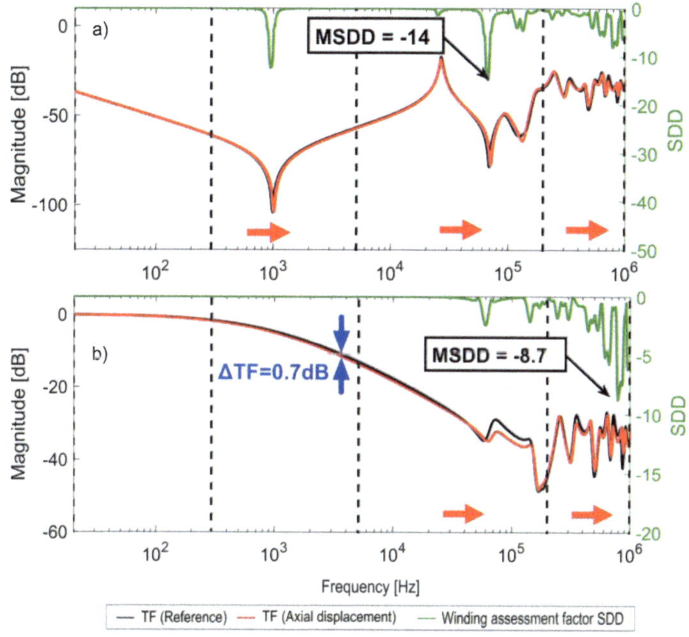

Figure 5-10: TFs of healthy and axially displaced B-phase winding (vector group YNyn0), and SDD as a measure of deviation between TFs; (a) HV EE-OC (b) HV EE-SC

In Figure 5-10, the calculated *SDD* is also plotted as a winding assessment factor which gives a quantitative measure of deviation between TFs. The minimum value of *SDD (MSDD)* indicates the maximum deviation between TFs. It can be seen that TFs of inter-winding configurations (IIW and CIW) possess the lowest value of SDD indicating that these configurations are more sensitive to AI-M.

5.3.1.2 Hoop buckling (HB-M) and Compressive tension fault (CTF-M)

Mainly, two types of radial deformations can be found, i.e., compression failure in inner windings and hoop tension failure in outer windings. Generally, the term “buckling” is used for both types of radial deformations [5]. However, their effects on the FRA spectrum can be different in different cases. In the HF model, both modes of radial deformation are simulated as shown in Figure 5-11. For this purpose, one section of the B-phase winding is deformed throughout the winding height. Considering Figure 5-11, the radius of the radial deformation $r(\theta)$ is modeled as given:

$$r(\theta) = \begin{cases} r_o - \frac{d}{2}(\cos(s\theta - 1) & 0 \leq \theta \leq \frac{2\pi}{s} \\ r_o & otherwise \end{cases} \tag{5-2}$$

Where r_o is the non-deformed radius, d is the deformation depth (positive d: compression failure of LV winding and negative d: hoop tension failure of the HV winding), s is the span of radial deformation, θ is the arc angle. The arc angle of the deformed section is kept at 30° and the RDF level is calculated based on the depth of the deformed and the non-deformed radius of the winding as follows:

$$\%\, RD\, fault\, level = \frac{d}{r_0} \times 100\% \tag{5-3}$$

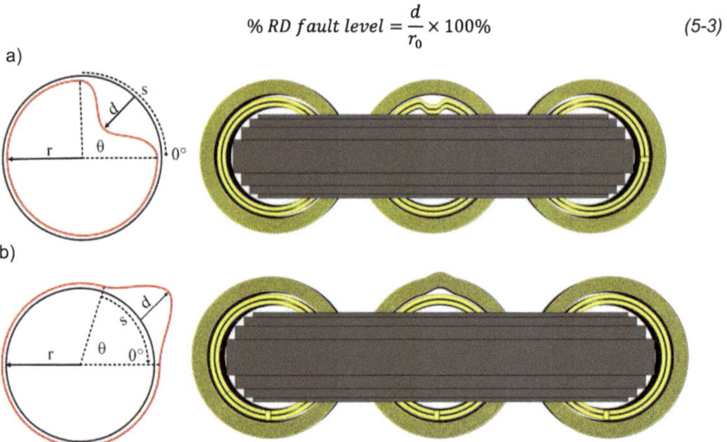

Figure 5-11: Representation of Radial Deformation Fault (RDF); schematic (left); 3D HF model (right) (a) compression failure in LV winding (b) hoop tension failure in HV winding

In the HF model, fault level of 10% and 5% is simulated in compression failure and hoop tension failure, respectively. The effects of both types of buckling on simulated FRA signatures for EE-OC and EE-SC connection schemes are shown in Figure 5-12 and Figure 5-13, respectively. While the results of interwinding connection schemes are presented in Appendix F. To identify the degree of deviation between TFs, *SDD* is also displayed on the same graph. Under buckling fault, the distance of the winding to the ground (C_g: capacitance to ground), the interaction between windings (C_w and M_u), the leakage inductance (L) and winding's series capacitances (C_s) are changed. Consequently, the effects of buckling fault are noticeable in the MFB and HFB. It can be noticed that the impact of both types of bucklings is quite similar, as both affect similar parameters. The impact of buckling fault in different connection schemes is briefly explained in the following:

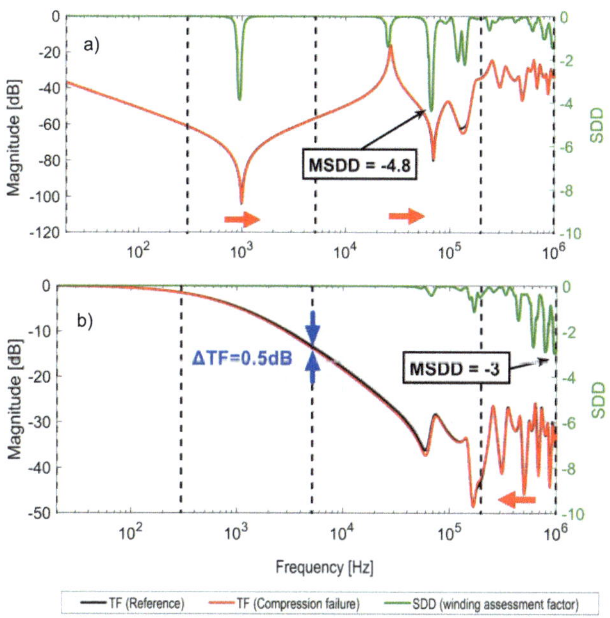

Figure 5-12: TFs of healthy and deformed B-phase LV winding (vector group YNyn0), SDD as a measure of deviation between TFs; (a) HV EE-OC (b) HV-SC

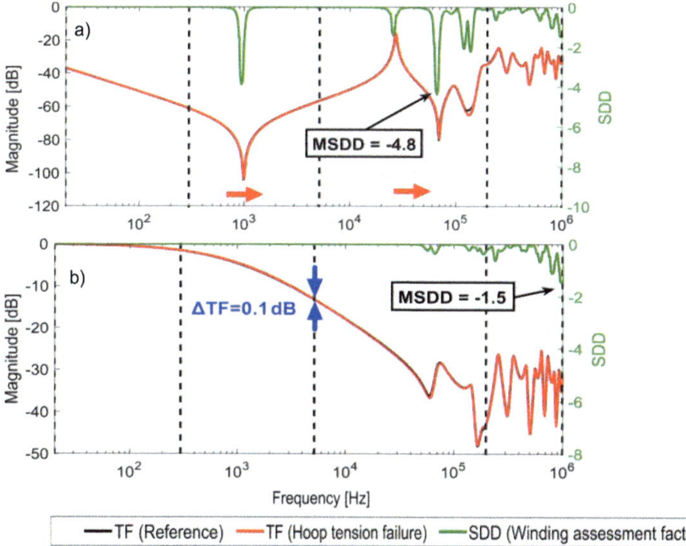

Figure 5-13: TFs of healthy and deformed B-phase HV winding (vector group YNyn0) and SDD as a measure of deviation between TFs; (a) HV EE-OC (b) HV EE-SC

HV EE-OC (Figure 5-12a and Figure 5-13a): For both types of buckling the LFB1 is unchanged. In LFB2, the first anti-resonance point is slightly shifted to the right (represented by the red arrows) due to the change of C_g, C_w, and C_s. The deviations are most obvious in the MFB where the resonance frequencies are shifted to the right due to the decrease of Ciw and Mu. Small deviations at resonance points were also found in HFB, as detected by *SDD*. These changes can be attributed to the change of C_s.

HV EE-SC (Figure 5-12b and Figure 5-13b): For both types of buckling, TFs are changed in the inductive roll-off region. In this region, due to the change in leakage inductance, a change in TFs is observed. The difference in magnitude is 0.5 dB and 0.1 dB for compression and hoop tension failure, respectively. In HFB, both faults result in slight frequency and magnitude shifts.

IIW (Figure F.3 and Figure F.4): TFs in LFB1 and LFB2 remained unaffected under both types of bucklings. Because RDF does not affect the turns ratio. In HFB TFs are slightly changed as indicated by *SDD*.

CIW (Figure F.2 and Figure F.4): The linear behavior of TFs in LFB1 and at the end of LFB2, is dominated by capacitance between windings (C_w). Due to the decrease of C_w, the TFs are shifted downward.

By comparing the values of *MSDD* in different connection schemes, EE-OC and CIW configurations can be identified as the most sensitive configurations to detect buckling faults. In IEEE std. C57 [8] and CIGRE-342-2008 [22], few case studies regarding compression failure in inner winding are reported. The simulation results of the HF model show similar trends.

5.3.1.3 Conductor tilting fault (CT-M)

In electromagnetically balanced windings, the axial forces act to axially compress the windings. When the axial compression force exceeds a certain limit, a fault called "conductor titling" occurs, which is a principal mode of failure in large power transformers [5]. In this mode of failure, the conductors turn around their axis of symmetry. The conductor tilting fault may result in damage to insulation paper, displacement of conductors, shorted-turn fault, and even collapse of the winding in severe cases. Figure 5-14 illustrates the CT-M in real transformer winding [7] and its equivalent simulation in the HF model. Conductor tilting is simulated by tilting the conductors in the entire circumference of the B-phase HV winding. The angle of tilt is kept at 15° whereas the maximum tilt angle is considered to be 90°. CT-M is applied to the bottom 30 disks, and the fault level is calculated based on the number of discs with tilted conductors as follows:

$$\%\ \text{CT-M fault} = \frac{\text{number of faulted discs}}{\text{number of total discs}} \times \frac{\text{applied tilt angle}}{\text{max possible tilt angle}} \times 100\% \qquad (5\text{-}4)$$

In this way, 8% of the CT-M is simulated in the HF model. The effects of CT-M on simulated FRA signatures for end-to-end connection schemes are shown in Figure 5-15. While the results of interwinding connection schemes are presented in Appendix F. To identify the degree of deviation between TFs, *SDD* is also displayed on the same graph. The impact of CT fault in different connection schemes is briefly explained in the following:

HV EE-OC: The effect mainly prevailed in HFB as indicated by *MSDD*. In this region, the resonance frequencies are shifted to the left (indicated by red arrows). The shift in the resonance can be attributed to the change of C_s of the winding. Because CT-M is caused by axial compression of discs, thus the separation between the conductors changes which eventually changes C_s.

HV EE-SC: Due to the change in impedance the TF of the affected winding exhibits an attenuation of 0.2 dB within the inductive roll-off region. In MFB the effect is small and difficult to identify. While in HFB largest differences in TFs are produced which is also evident from the minimum value of *SDD*.

IIW (Figure F.5): The differences between TFs are noticeable in HFB. While other regions remain unaffected.

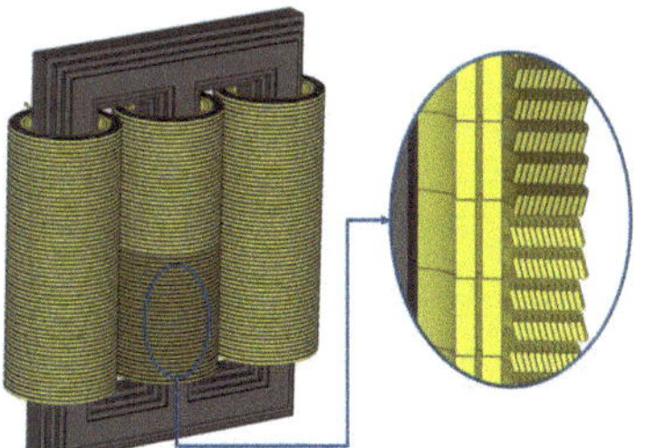

Figure 5-14: Representation of conductor tilting failure mode in the HF model (left) and 275/132-kV 240-MVA autotransformer [7] (right)

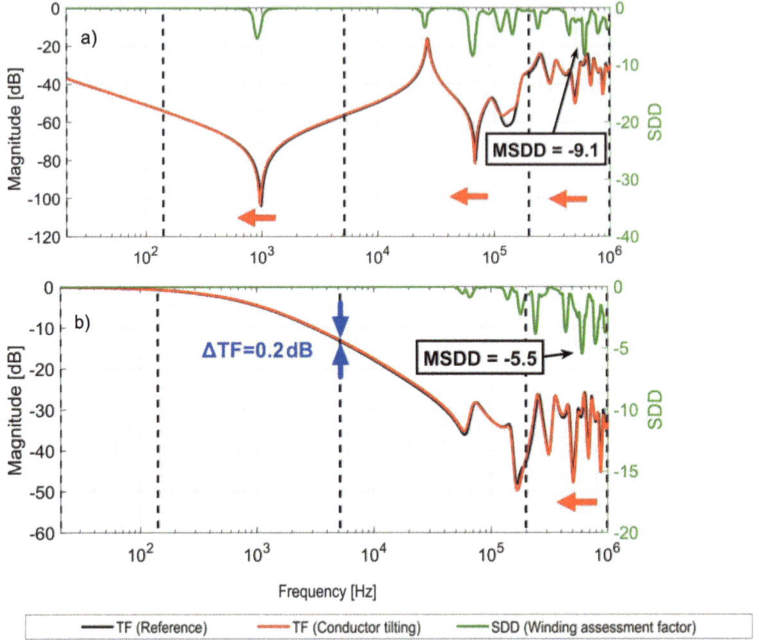

Figure 5-15: TFs of healthy and deformed (conductor tilting failure) B-phase HV winding and SDD as a measure of deviation between TFs; (a) HV EE-OC (b) HV EE-SC

CIW (Figure F.5): In the linear behavior of TFs in LFB1 and LFB2, the TF of the affected winding is shifted upward. This change can be attributed to the change in the C_s and C_w. Whereas in HFB, the resonance peaks and valleys are shifted which is a key indicator of the dominant effect of C_s in this region.

Looking at the values of *MSDD* in different connection schemes, CIW proved to be the most sensitive connection scheme to detect the CT-M. Very few studies are found in the literature as CT-M fault is difficult to implement. In CIGRE-342-2008, one case study is reported [7]. The simulation results of the HF model show similar trends against conductor tilting failure mode.

5.3.1.4 Axial disk buckling fault (ADBF) (CB-M)

The opposing axial forces on the windings can also bend the conductors between insulating spacers which are located radially or leaned to each other. The occurrence of this phenomenon is called bending or axial disk buckling. A schematic representation of the ADBF, in a typical transformer winding, is shown in Figure

5-16a. In the HF model, this fault is simulated by introducing a vertical buckling to the bottom 30 disks in HV winding as shown in Figure 5-16b. The buckling distance (h) was set to 4 mm. While the distance (d) between normal disks was 5 mm. The arc angle of the deformed section is kept at 75°. The percentage ADBF level was calculated as follows:

$$\%\ \text{ADB fault} = \frac{\text{number of faulted discs}}{\text{number of total discs}} \times \frac{\text{deformed arc angle}}{360°} \times \frac{\text{h}}{\text{d}} \times 100\% \quad (5\text{-}5)$$

To study the impact of ADBF on TFs of different connection schemes, an 8% fault level is implemented in the HF model. The simulated TFs along with the application of the winding assessment factor *(SDD)* is presented in Figure 5-17. The effects of ADBF in each connection scheme are briefly explained in the following:

HV EE-OC: Results do not lead to drastic deviations in LFBs and MFB as the value of *SDD* approaches to zero in these bands. ADBF only affects the winding structure. Consequently, the effects only appear in HFB where the resonance frequencies are shifted to the left (indicated by the red arrow). These effects are attributed to the changes in the inter-disk capacitance and M_u between the affected disks. The maximum change between TFs in HFB is also indicated by the minimum value of *SDD*.

HV EE-SC: The TFs of the affected case are offset in the inductive roll-off portion. In this region, due to the change of the impedance, the difference in magnitude is greater than 0.2 dB. In HFB, the effect is similar to the HV EE-OC.

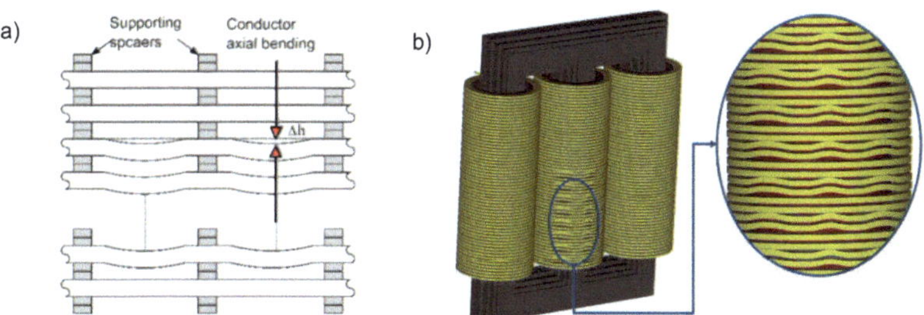

Figure 5-16: Representation of axial disk buckling failure mode; (a) schematic (b) B-phase HV winding of 3D HF model

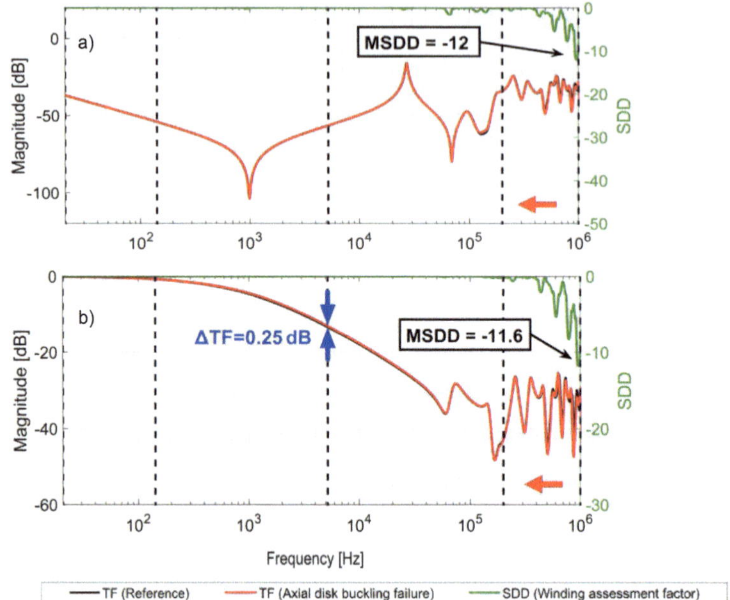

Figure 5-17: TFs of healthy and disk buckling failure in B-phase HV winding and SDD as a measure of deviation between TFs; (a) HV EE-OC (b) HV EE-SC

IIW (Figure F.6): The TF of the affected winding is shifted to the left in HFB, recognized by the minimum value of *SDD*. While other frequency bands are unchanged.

CIW (Figure F.7): In the linear behavior of TFs in LFB1 and LFB2, the TF of the affected winding is slightly attenuated but this change is minimal and difficult to identify. Whereas in HFB, the resonance peaks and valleys are shifted which are also witnessed by the minimum value of *SDD*.

Looking at the values of *MSDD* in different connection schemes, CIW can be identified as the most sensitive connection scheme to detect the ADBF.

5.3.2 Electrical Failure Modes

5.3.2.1 Short Circuit between Conductors (SCC-E)

A short circuit between conductors is the most common electrical failure mode in transformer windings. Depending upon the severity it can influence the characteristics of magnetizing inductance, winding resistance, and self-inductance of the windings. Shorted turn fault gives rise to large circulating currents which leads to localized thermal overloading, thereby causing hot spots. With time, the fault manifests itself and may lead to catastrophic failure of the transformer. In the HF model, SCC-E is simulated by shortening two turns at various locations in the B-phase HV winding. It was found that the shorted-turn fault in the middle of the winding has a greater impact on the TFs of the affected winding. This can be associated with the higher mutual inductance of the middle disks in comparison to the top and bottom disks. For this reason, only the results of one SC location (middle) are included in this paper. When a SCC-E fault occurs, a large amount of flux passes through the air instead of the core and surrounds the shorted-turn. Thus, the leakage flux is essentially increased at the fault location. Consequently, the magnetic reluctance and self-inductance of the corresponding limb changes. Figure 5-18 shows the impact of SCC-E on simulated TFs of different connection schemes. To identify the degree of deviation between TFs, *SDD* is also displayed on the same graph. The impact of SCC-E in different connection schemes is briefly explained in the following:

HV EE-OC: SCC-E is most obvious in the low-frequency region. As stated earlier that SCC-E remarkably changes the magnetizing characteristics of the core and results in reduced magnetizing inductance. Consequently, the TF of the affected winding is offset in LFBs and the first resonance frequency is shifted up to 250% in comparison with the healthy case. Slight deviations can also be seen in the high-frequency region due to the change in the impedance of the winding.

HV EE-SC: The inductive roll-off region is affected due to the change of leakage inductance. In this region, a decrease in impedance is observed and the difference in magnitude is greater than 0.2 dB. This region provides the possibility of a quantitative diagnosis of SCC-E, and a diagnosis criterion can be set to a permissible change of the impedance. The low-frequency response (inductive roll-off portion) of short circuit configuration is equivalent to short circuit impedance. According to IEC-2200-2014 [35], if the short-circuit impedance of a transformer changes by 2% or more then the transformer is recognized as damaged.

IIW (Figure F.8): TFs in LFB1 and LFB2 are slightly offset due to the change in the number of turns. But this change is minimal and difficult to identify especially if fewer turns are shorted. In MFB and HFB, the resonance points are shifted and new resonance peaks and valleys are produced.

CIW (Figure F.8): In LFB2, the first resonance frequency is affected. In MFB and HFB, the resonance frequencies are shifted to the right due to the decrease in the impedance of the winding.

Remarkable changes in the TFs of EE-OC and CIW configurations indicate that these connection schemes are more sensitive to detect SCC-E. Various studies related to the diagnosis of windings short circuit faults can be found in the literature. The presented simulation results of SCC-E aligns well with the practical results reported in [5], [7], [11].

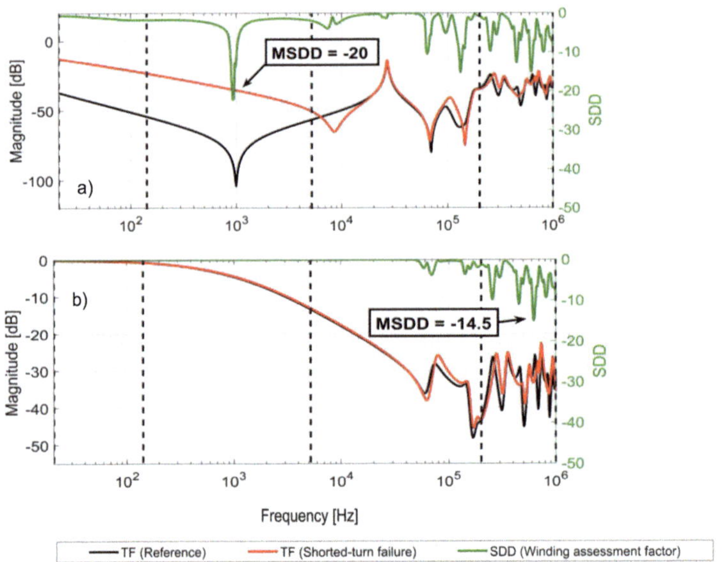

Figure 5-18: TFs of healthy and SCC-E failure in B-phase HV winding and SDD as a measure of deviation between TFs; (a) HV EE-OC (b) HV EE-SC

5.3.2.2 Core Ground Loss (CGL-E)

Mostly, the connection to the ground of the core is made internally, thus making it difficult to simulate and understand the effect of an ungrounded core. In the HF model, it is possible to disconnect the core connection to the ground to study the impact of the missing core ground on TFs of different connection schemes.

Generally, the CGL-E changes the LV winding-to-ground capacitance which results in the shifting of HV winding response in the medium and high-frequency range. Figure 5-19 shows the impact of CGL-E on simulated TFs of different connection schemes. To identify the degree of deviation between TFs, *SDD* is also displayed on the same graph. The impact of CGL-E in different connection schemes is briefly explained in the following:

HV EE-OC: The effect of CGL-E is most obvious in the MFB and HFB. As mentioned earlier, CGL-E results in additional capacitive current flowing from LV winding and inter-winding capacitances towards the measuring impedance. This capacitive component will change the TF of the HV winding in MFB and HFB. *SDD* also detects the larger differences between TFs in these regions.

HV EE-SC: The effect of CGL-E is most prominent in the HFB where the resonance points are shifted and new peaks and valleys are produced.

IIW (Figure F.9): This configuration is the least sensitive for CGL-E. Slight differences between TFs appear in HFB. However, these differences are minimal and difficult to identify. As indicated by the minimum value of *SDD* (*MSDD*=-0.4).

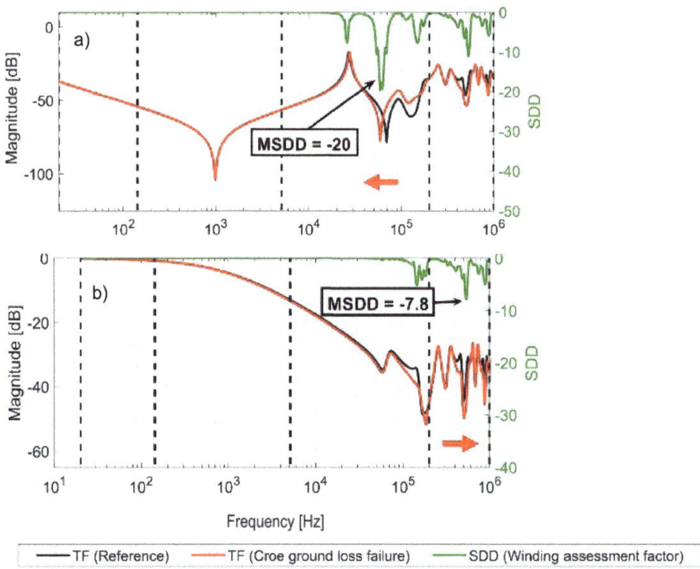

Figure 5-19: TFs of B-phase HV winding with and without CGL-E (vector group YNyn0) and SDD as a measure of deviation between TFs; (a) HV EE-OC (b) HV EE-SC

CIW (Figure F.9): At low frequency, inter-winding capacitance dominates the response and a very slight deviation exists due to the change of ground capacitance that affects the inter-winding capacitance. The influence of the missing core is increased at high frequencies around 1 MHz.

From Figure 5-19, the lowest value of *MSDD* in EE-OC indicates the higher sensitivity of this configuration. It was also observed that this failure mode affects the TFs of all the phases.

5.3.3 Sensitivity of connection schemes against failure modes

The sensitivities of FRA in various connection schemes against various faults are also an important issue and these are not fully investigated yet. In the field, mostly end-to-end measurements are performed and inter-winding measurements are skipped. In literature, only a few studies can be found to objectify the most sensitive FRA scheme for a few mechanical faults. The application of the proposed HF model gives the possibility to investigate the most sensitive connection schemes for different electrical and mechanical failure modes.

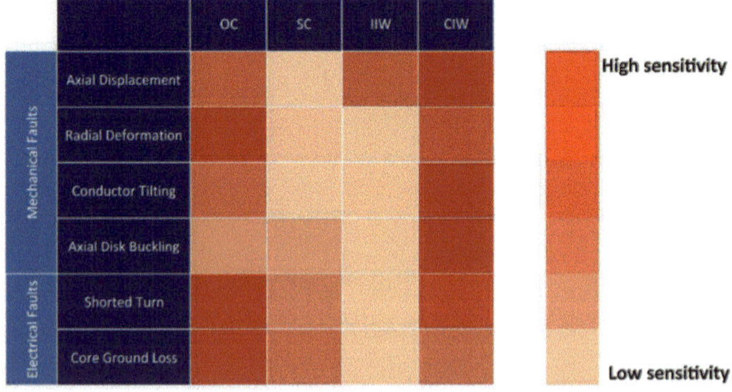

Figure 5-20: Identification of most sensitive connection scheme for various electrical and mechanical faults; patterns are based on the minimum value of SDD (MSDD)

The most sensitive connection scheme against each investigated fault presented in the previous sections can be identified from Figure 5-20[1]. Here sensitivity pat-

[1] *The values of MSDD are normalized for each fault type.*

terns are based on the minimum value of the *SDD (MSDD)* in different configurations under different investigated faults. It can be noticed that the effect of different faults has differently prevailed in all four connection schemes. It can be seen that EE-OC configuration exhibits the highest sensitivity against radial deformation, shorted turn, and missing core ground faults. While CIW configuration holds the highest sensitivity to detect axial displacement, conductor tilting, axial disk buckling, and shorted turn faults.

In summary, EE-OC and CIW connection schemes have proved to be more sensitive under various investigated electrical and mechanical faults. It is noteworthy to mention that derived sensitivity also depends on the definition of the applied indicator (*SDD*) and may change with the application of different indicators.

5.3.4 Deviation patterns of electrical and mechanical faults

Figure 5-21[2] summarises the characteristic impact of individual faults on the TFs of four connection schemes, i.e., EE-OC, EE-SC, CIW, and IIW. These deviation patterns are based on the minimum value of *SDD (MSDD)* in different frequency sub-bands for each fault type.

Based on the presented case studies of mechanical and electrical failure modes, it can be observed that different patterns of deviations can be characterized for different faults according to the frequency sub-bands. Consequently, it is possible to classify different faults based on their deviation patterns in various frequency sub-bands. Thus, it can be summarized from the above case studies that the *SDD* indicator has the ability to detect and identify the fault types using the proposed frequency sub-band division structure. These deviation patterns can be served as features to train and develop intelligent fault detection and classification algorithm.

The results of this study can be used as a way forward for the establishment of a standard algorithm for power transformer fault detection and classification. By embedding the frequency division algorithm along with the winding assessment factor (*SDD*) into the FRA instruments, an automatic condition assessment of transformers could be realized for FRA measurements.

---

[2] *Absolute value of MSDD is employed for heat chart of deviation patterens*

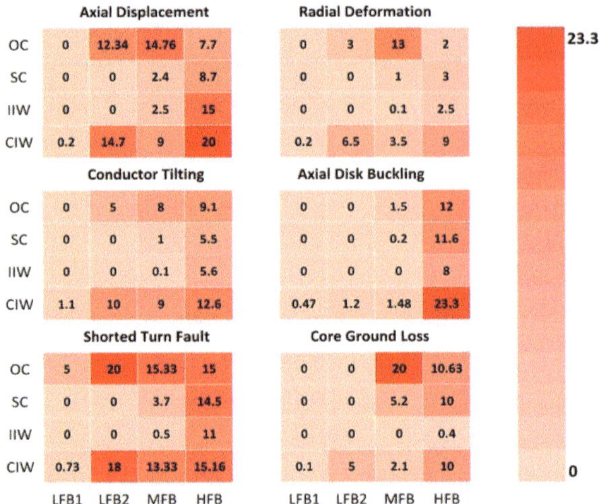

Figure 5-21: Deviation patterns based on the minimum value of SDD (MSDD) for fault classification

5.3.5 Summary

The proposed HF model eliminates the demand for parameter estimation which is required in lumped parameter models. In this model precise and accurate fault simulations are possible. This feature enables the transformer manufacturers to use the design files of the transformer in the CST MW Studio, simulate several faults, and evaluate the winding assessment factor (*SDD*). Consequently, it is advisable to set different thresholds for various transformers and this information can be delivered to the customers as a characteristic of the power transformer.

The proposed HF model can be used to generate a large database that covers all the possible winding faults. This database can be used to generate different deviation patterns of the *SDD* indicator using the proposed frequency sub-band division structure. To develop an automatic condition assessment tool, these deviation patterns can be served as features to train and develop an intelligent fault detection and classification algorithm, i.e., artificial neural network. Additionally, the proposed model is also useful to study the effects of different electrical properties (moisture effect, temperature effect, etc.) of windings on the frequency response. Besides, the proposed model is also useful to study the transient analysis of power transformers and the calculation of winding overvoltage under different fault conditions.

6 Transformer equivalent circuit parameter estimation

This chapter presents a mathematical approach to extract transformer equivalent circuit parameters from FRA measurements. These extracted parameters are then used to assess the condition of the transformer. The performance of the method is analyzed with case studies.

6.1 Principle of Transformer Parameter Estimation

The active part of a power transformer forms a complex RLC network that can be represented by an equivalent circuit as shown in Figure 6-1. Thereby, the frequency response of a transformer has a fundamental relationship with the core, the interaction between windings, and the structure of the windings. As appreciated by IEC and IEEE standards [5], [11], mainly four test configurations are recommended, i.e., EE-OC, EE-SC, CIW, and IIW. The frequency response for each configuration gives information about some specific physical part of the transformer. Hence, the transformer equivalent circuit parameters can be extracted from the frequency responses of all four configurations. However, a first approach is essential for differentiating the dominant effects of different parameters in different configurations. A typical example of a transformer FRA measurements for four connection schemes are shown in Figure 6-2 and features of these configurations are described in the following:

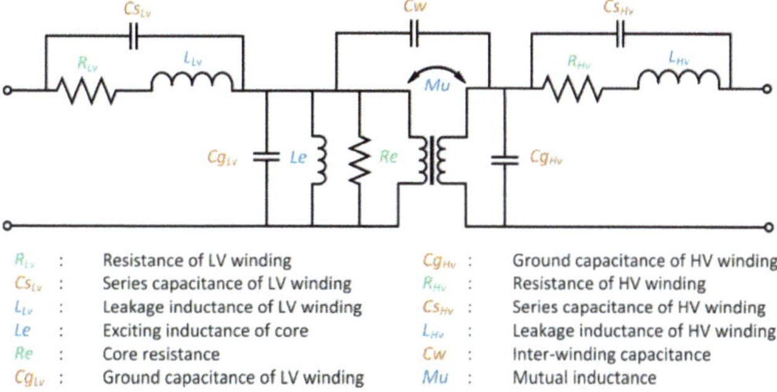

Figure 6-1: RLC parameters of transformer equivalent circuit

EE-OC: At low frequencies, the response is dominated by core magnetizing inductance, and shows a decreasing trend of 20 dB/decade. With the increase of frequency, the winding capacitances start to exhibit resonant coupling effects, thus, producing anti-resonances. In the medium frequency range, the response is dominated by the interaction between the windings. At high frequency, the response is mainly dominated by the winding capacitances and inductances [5].

EE-SC: At low frequencies, the response is characterized by the leakage inductance. At higher frequencies, the response is similar to the open-circuit configuration [5].

CIW: The response of this configuration is capacitive in nature. At low frequencies, the response is linear and dominated by the inter-winding capacitance [5].

IIW: At low frequencies, the response is characterized by the winding turns-ratio [5].

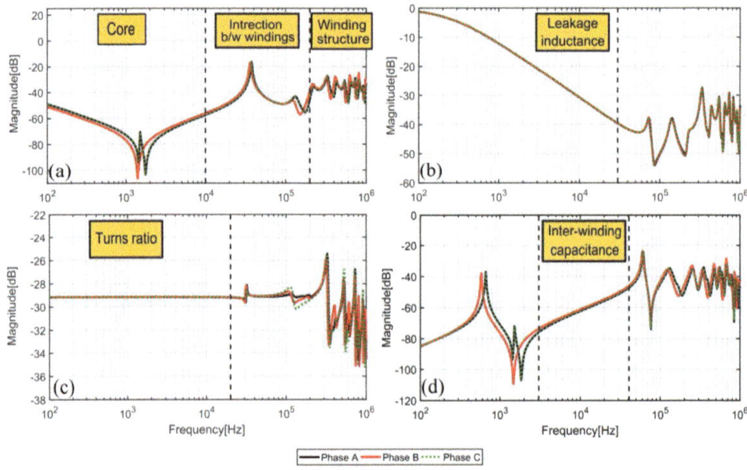

Figure 6-2: Frequency response of HF model of the three-phase transformer (vector group: YNyn0) for different test configurations; (a) EE-OC (b) EE-SC (c) IIW (d) CIW

6.2 Sensitivity Analysis

The effect of exciting (leakage) inductances in EE-OC (EE-SC) FRA signature is quite clear, as many studies have investigated their response [39], [95], [96], [97]. However, the effect of capacitances is not fully explained in the low and medium frequency regions. Heindl et at. [2010] [95], suggest that the first anti-resonance is determined by the series capacitance (C_s) and exciting inductance (L_e). While Gonzales et al. [2012] [96], state that the first anti-resonance is due to the coupling between ground capacitances (C_g) of windings and exciting inductance (L_e) of the core. Cheng et al. [2019] [97] suggest that the first anti-resonance is influenced by the network capacitance ($C_s + C_g + C_w$).

In this research, a sensitivity study on capacitive parameters (C_s, C_g *and* C_w) is performed to identify their effects on FRA signatures. For this purpose, the HF 3-phase transformer simulation model (as presented in Chapter 5) is employed. The reference values for these three capacitances are shown in Table 6-1. For convenience, the TFs of V-phase winding are chosen as an example. Three capacitive parameters of the equivalent circuit, i.e., C_g, C_s, and C_w are increased by 45% of their original values, and the TFs of EE-OC and CIW configurations are plotted for sensitivity analysis. It should be noted that capacitive parameters of all three phases are changed, also the percentage frequency shifts at resonance and anti-resonance points are indicated for each parameter change as shown in Figure 6-3 and Figure 6-4.

Table 6-1: Reference values of capacitive parameters of three-phase transformer (phase-V)

Parameters	C_w	C_{gLv}	C_{gHv}	C_{sLv}	C_{sHv}
Values	587 pF	527 pF	113 pF	1115 pF	765 pF

Figure 6-3 shows the EE-OC TFs of V-phase winding before and after the change of capacitive parameters. From Figure 6-3(a) and Figure 6-3(b), the impact of ground capacitances (C_{gLv} and C_{gHv}) can be analyzed. Both affect the low-frequency region at the first anti-resonance frequency, shifting it to the left with the increase of C_g. However, the effect of C_{gLv} is more prevailed (Δf=-14%). The impact is also noticeable in the medium and high-frequency region, again shifting the resonance frequencies to the left. Figure 6-3(c) illustrates that C_w affects the first anti-resonance in the low-frequency region and series resonance in the medium frequency where the resonance frequencies are moved to the left. The effect of C_{sHv} is presented in Figure 6-3(d). It can be seen that C_{sHv} has a negligible effect at the first anti-resonance point. The effect is prominent only at high frequencies, i.e. above 800 kHz.

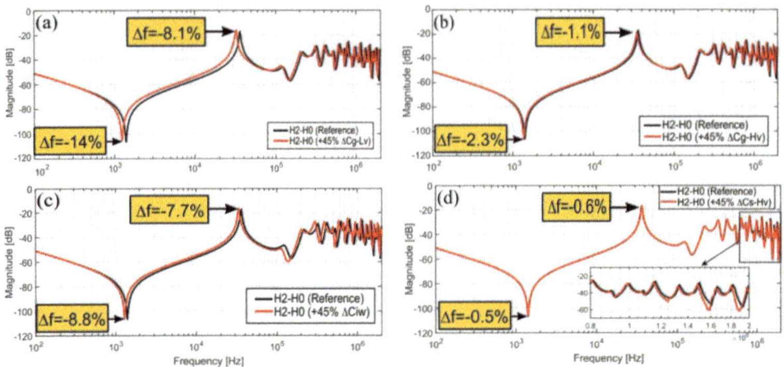

Figure 6-3: Sensitivity study of capacitive parameters on EE-OC TFs; (a)ΔC_{gLv} (b)ΔC_{gHv} (c)ΔC_w (d)ΔC_{sHv}

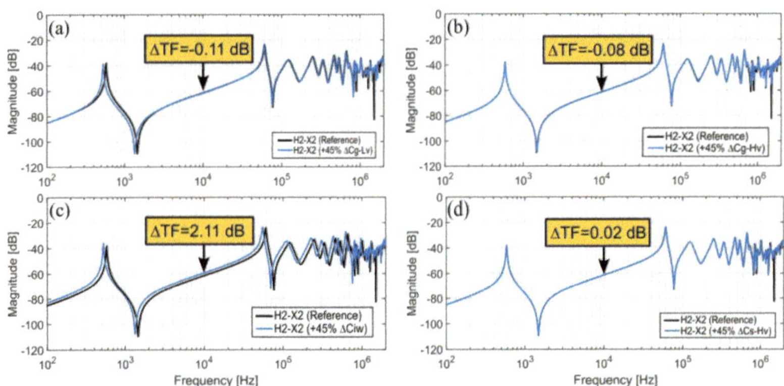

Figure 6-4: Sensitivity study of capacitive parameters on CIW TFs (a) ΔC_{gLv} (b) ΔC_{gHv} (c) ΔC_w (d) ΔC_{sHv}

From these results, it can be concluded that the 1st anti-resonance of EE-OC TF is produced by the coupling between the core inductance and the equivalent capacitance (C_{eq}) which is a combination of network capacitances (C_{gLv}, C_{gHv}v, and C_w). However, C_g is the most dominant factor. It should be noted that the increasing capacitive trend after the first anti-resonance point is also affected by C_{eq}. Figure 6-4 shows the CIW TFs of V-phase winding before and after the change of capacitive parameters. As mentioned before, the response of this configuration shows an increasing capacitive trend just after the first anti-resonance, characterized by C_w. It is also evident from Figure 6-4, that only the effect of ΔC_w is prominent in this region. While the effect of other network capacitances is negligible in this region. The change in TF values due to changes in capacitive parameters is also indicated in Figure 6-4.

6.3 Development of interpretation methodology

As discussed above, the frequency response for each configuration gives information about some specific physical parts of the transformer. Hence, the transformer equivalent circuit parameters can be estimated from the frequency responses of all four configurations. The methodology of parameter estimation is based on the fact that in FRA, the frequency response of the transformer impedance is measured which is characterized by the RLC parameters of the windings. The magnitude MAG of the frequency response can be related to the impedance of the transformer $Z(\omega)$, using Equations (6-1) - (6-4).

$$MAG(\omega) = 20\log\left(\frac{50}{50 + Z(\omega)}\right) \tag{6-1}$$

$$\mathrm{Re}\{Z(\omega)\} = \left(\frac{50}{10^{MAG/20}} - 50\right) \times \cos\varphi_{MAG} \tag{6-2}$$

$$\mathrm{Im}\{Z(\omega)\} = \left(\frac{50}{10^{MAG/20}} - 50\right) \times \sin\varphi_{MAG} \tag{6-3}$$

$$|Z(\omega)| = \sqrt{(\mathrm{Re}\{Z(\omega)\})^2 + (\mathrm{Im}\{Z(\omega)\})^2}, \quad \varphi_Z = \tan^{-1}\left(-\frac{\mathrm{Im}\{Z(\omega)\}}{\mathrm{Re}\{Z(\omega)\}}\right) \tag{6-4}$$

The frequency response of a transformer is usually represented by magnitude MAG and phase plots φ. Thus, at any frequency, the impedance of the transformer can be determined from MAG and φ using Equations (6-2) and (6-3). The details of parameter extraction from each test configuration are given below:

6.3.1 Parameter estimation from EE-OC configuration

The TF of EE-OC configuration is influenced by the core properties at low frequency, the interaction between windings at medium frequency, and the winding structure at high frequency. Therefore, exciting inductance *(L_e)*, core resistance *(R_e)*, mutual inductance *(M_u)*, and equivalent capacitance *(C_{eq})* can be extracted from this configuration. The equivalent magnetic circuit is represented in Figure

6-5. The TF of EE-OC configuration can be seen as a parallel circuit of L_e and R_e as shown in Figure 6-5(b), in the frequency band lower than the first anti-resonance frequency which is the region I in the case of Figure 6-6. Therefore, R_e and L_e can be determined from real and imaginary parts of the impedance *(Z)* in region I. Due to different magnetic paths, the lateral phases exhibit slightly lower L_e.

In region II of Figure 6-6, TF can be seen as two parallel circuits of L_e, R_e, and C_{eq} connected in series as illustrated in Figure 6-5(c). Generally, in a three-phase transformer, the lateral phases are characterized by two anti-resonances. Thus, for the lateral phases, L_e can be represented as a combination of two exciting inductances (L_{e1} *and* L_{e2}). This is due to different magnetic paths as shown in Figure 6-5(a). The average path length of the central phase is lower than that of the lateral phases. So, the reluctance of the central phase will be less than that of the lateral phases, and its inductance *(*L_{e1}*)* will be higher. Thus, the first anti-resonance frequency corresponds to *Le1* while the second anti-resonance belongs to L_{e2}. Both L_{e1} and L_{e2} can be estimated from equations (6-5) and (6-6), respectively. Where $f_{parallel-res1}$ and $f_{parallel-res2}$ are the first and second anti-resonance frequencies, respectively. $f_{series-res1}$ is the series resonance, C_{eq} is the equivalent capacitance and M_u is the mutual inductance.

$$f_{parallel-res1} = \frac{1}{2\pi\sqrt{L_{e1}C_{eq}}} \quad \text{(6-5)}$$

$$f_{parallel-res2} = \frac{1}{2\pi\sqrt{L_{e2}C_{eq}}} \quad \text{(6-6)}$$

$$f_{series-res1} = \frac{1}{2\pi\sqrt{M_u C_{eq}}} \quad \text{(6-7)}$$

While C_{eq} can be estimated from the increasing capacitive trend of region II as illustrated in Figure 6-6. As stated above, that the frequency response after the first anti-resonance is followed by the increasing capacitive trend which is sensitive to the C_{eq}. Thus, C_{eq} can be estimated in this region using Equation (6-3), at the frequency where the phase approaches 90°. As mentioned before that medium frequency region belongs to the interaction between windings. This region is characterized by one or two series resonances depending upon the type of the transformer. For example, the frequency response of autotransformers possesses two series resonances for both series and common windings. This behavior is often termed as a double-peak feature. At these resonance frequencies, the TF of EE-OC configuration can be seen as a series combination of C_{eq} and M_u. Thus, knowing C_{eq} the value of M_u can be estimated using Equation (6-7).

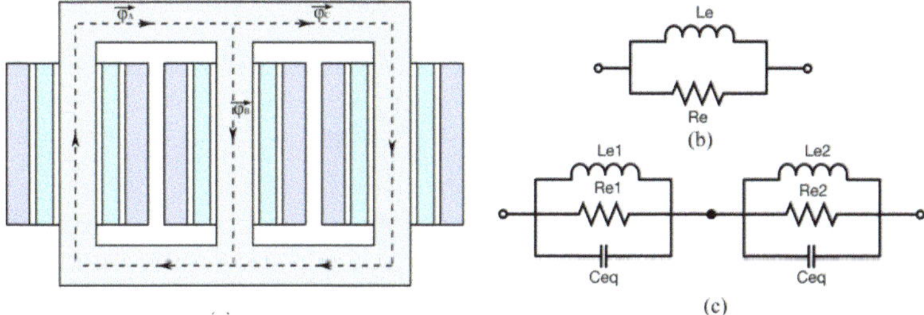

Figure 6-5: Representation of equivalent circuits;

(a) magnetic circuit

(b) equivalent RL circuit for the low-frequency region of EE-OC TF

(c) equivalent RLC circuit for the first two parallel resonance points in EE-OC TF

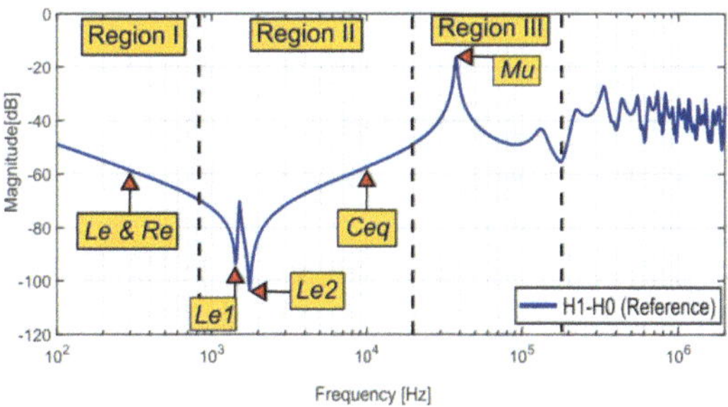

Figure 6-6: Equivalent circuit parameter estimation from EE-OC TFs

6.3.2 Parameter estimation from EE-SC configuration

EE-SC configuration allows the influence of the core to be reduced at lower frequencies. The first anti-resonance frequency is increased because the inductance of the transformer with one winding short-circuited is significantly reduced. In this configuration, the first anti-resonance is caused by leakage inductance *(L_l)* instead of core exciting inductance. Consequently, the TF in EE-SC configuration of HV-windings can be seen as a series circuit of leakage inductances *(L_{Lv} and L_{Hv})* and winding resistances *(R_{Lv} and R_{Hv})* at low-frequency where the TF exhibits the constant slope. In the case of Figure 6-7, this frequency region is lower than about 30 kHz. Therefore, L_l *(L_{Lv} + L_{Hv})* and R_w *(R_{Lv} + R_{Hv})* can be determined from the real and imaginary part of the impedance *(Z)* in the region I using Equations (6-2) and (6-3), respectively.

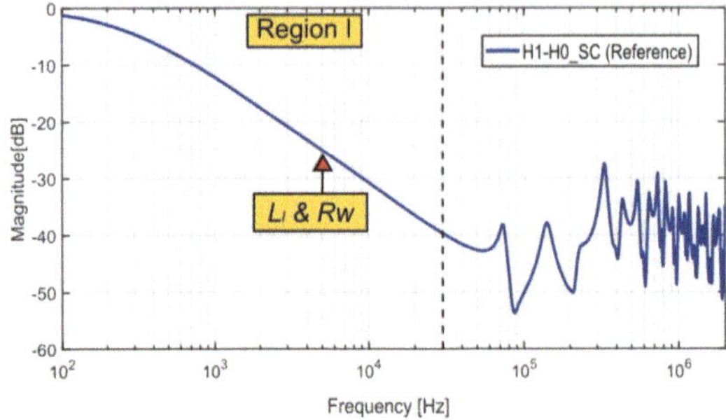

Figure 6-7: Equivalent circuit parameter estimation from EE-SC TF

6.3.3 Parameter estimation from inter-winding configurations

As described above, the TF of CIW configuration shows a linear behavior that is dominated by C_w, in the low-frequency region in which the TF exhibits the constant slope. In the case of Figure 6-8, this frequency band is indicated by Region I. In this frequency range (at the frequency where phase approaches 90°), C_w can be determined using the imaginary part of Z from Equation (6-3).

The inductive inter-winding measurement is performed to measure the voltage ratio between two windings, also known as the transfer voltage measurement [5]. In this configuration, the TF is dominated by a constant magnitude response at the low frequency. In the case of Figure 6-8, this frequency region is lower than about 20 kHz. The TF in this range closely resembles the transformer turns-ratio *(N)* with a variation of 0.2% [5]. Thus, N can be calculated in this range using Equation (6-8).

$$N = \frac{1}{10^{MAG/_{20}}} \tag{6-8}$$

In this way, 10 parameters of the transformer equivalent circuit can be estimated from four configurations of both fingerprint and current FRA results as shown in Table 6-2. After parameter estimation, it is possible to compute the percentage change of these parameters after fault occurrence that can be employed as identifiers to diagnose faults in a transformer. In the next section, two case studies are included where different faults are diagnosed using the proposed method.

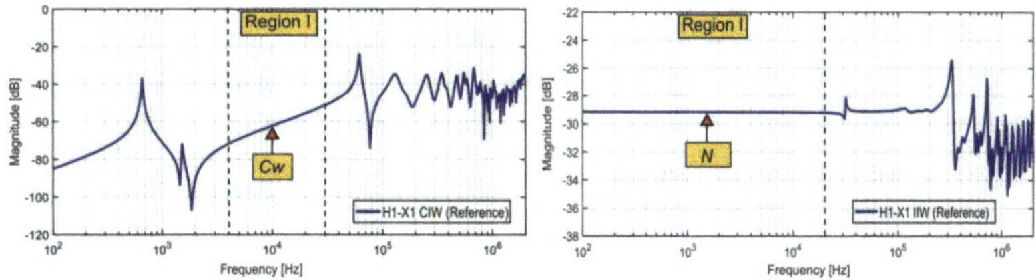

Figure 6-8: Equivalent circuit parameter estimation from CIW TFs (left) and IIW TFs (right)

Table 6-2: Estimated parameters from TFs of different configurations (Figure 6-6 – 6-8)

Connection schemes	EE-OC						EE-SC		CIW	IIW
Parameters	*Le*	*Re*	*Le1*	*Le2*	*Ceq*	*Mu*	L_l	*Rw*	*Cw*	*N*
Values	13.2 H	34.8 kΩ	28.8 H	19 H	427 pF	41.7 mH	26 mH	97.8 Ω	245 pF	28.7

6.4 Case Studies

In this section, two case studies (axial displacement and Radial deformation) are selected from the 3-phase HF transformer FRA model as described in Chapter 5.

6.4.1 Axial displacement fault

As described in Chapter 5, section 5.3, the axial displacement fault is simulated in the HV winding of the B-Phase. The HV winding is displaced downward by Δh=30 mm (~3.5% of its height). Figure 6-9 shows the effect of AD on the TFs of four connection schemes. The percentage change of estimated parameters due to AD fault is depicted in Table 6-3. The impact of AD fault in different connection schemes is briefly explained in the following:

EE-OC: TF remains unchanged in the low-frequency region. The first anti-resonance point is slightly shifted due to the change of the C_{eq}. The effect is most obvious in the medium (10–500 kHz) and high (500-1000 kHz) frequencies, where the resonance frequencies are shifted to the right due to the change of M_u and C_{eq}. This fact can also be perceived by the calculated percentage change of the estimated parameters, i.e., C_{eq} (-4.63%) and M_u (-17.6%) from Table 6-3.

EE-SC: The effect of AD fault can be perceived in the leakage region (typically below 1st anti-resonance frequency). In this region due to the change of impedance, the difference in TFs of healthy and affected winding is ~0.75 dB. At 8 kHz, the calculated percentage change of L_l is 10.3% as shown in Table 6-3.

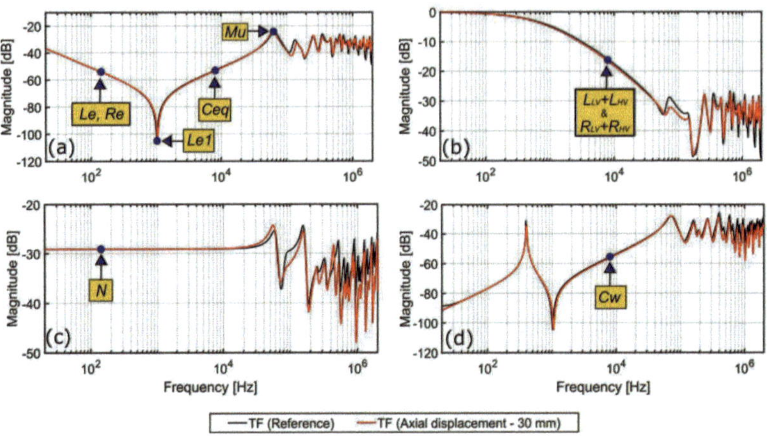

Figure 6-9: TF of healthy and displaced B-phase winding (vector group: YNd11); (a)EE-OC (b)EE-SC (c)IIW (d)CIW

IIW: TFs at low frequency (typically below 1 kHz) give a measure of the turns-ratio *(N)*, and it remained unaffected under AD fault. As can be seen from Table 6-3, *N* is unchanged.

CIW: Due to the decrease of C_w, the TFs in the linear increasing region (2-20 kHz) are shifted downward with a change of 0.6 dB. This is also verified by the calculated percentage change of C_w (-6.5%) as shown in Table 6-3.

Table 6-3: Percentage change of estimated parameters from TFs of different configurations

Connection schemes	EE-OC					EE-SC		CIW	IIW
Parameters	ΔLe (%)	ΔRe (%)	$\Delta Le1$ (%)	ΔCeq (%)	ΔMu (%)	ΔL_l (%)	ΔRw (%)	ΔCw (%)	ΔN (%)
Values	-0.08	6	0.14	-4.63	-17.6	10.3	8.4	-6.5	0.0

6.4.2 Radial deformation fault

Radial deformation fault is implemented in LV winding of B-phase as described in chapter 5, section 5.3. This type of fault is also known as compression failure in

LV winding or buckling in general. Figure 6-10 shows the effect of RD on TFs of four connection schemes. The percentage change of estimated parameters due to RD fault is depicted in Table 6-4. The impact of RD fault in different connection schemes is briefly explained in the following:

EE-OC: TF of the deformed case remains unchanged in the low-frequency region. The first anti-resonance point is slightly shifted due to the decrease of the C_w. The effect of RD fault is most obvious in the medium frequency region (10–500 kHz), where the resonance frequencies are shifted to the right due to the change of M_u and C_w. This fact can also be perceived by the calculated percentage change of C_{eq} (-1.3%) and M_u (-4.5%) as shown in Table 6-4.

EE-SC: The TF of the deformed case is changed in the leakage region. In this region, due to the change of leakage inductance, the difference in magnitude is 1.1 dB. The calculated percentage change of L_l is 16.1% as shown in Table 6-4.

IIW: TFs at low frequency give a measure of the N, and it remained unaffected under RD fault. As can be seen from Table 6-4, *N* is unchanged.

CIW: The linear increasing trend in the frequency region 2-20 kHz, is dominated by the capacitance between windings *(C_w)*. Due to the decrease of C_w, the TFs are shifted downward. This is also verified by the calculated percentage change of C_w (-8.2%) as shown in Table 6-4.

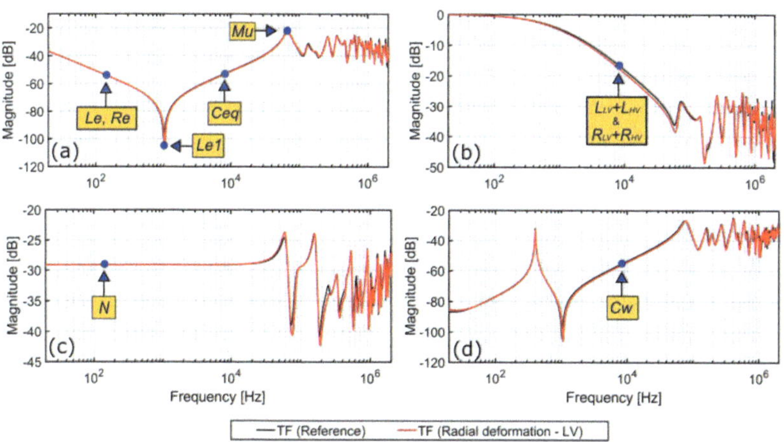

Figure 6-10: TF of healthy and displaced B-phase winding; (a)EE-OC (b)EE-SC (c)IIW (d)CIW

Table 6-4: Percentage change of estimated parameters from TFs of different configurations

Connection schemes	EE-OC					EE-SC		CIW	IIW
Parameters	ΔLe (%)	ΔRe (%)	ΔLe1 (%)	ΔCeq (%)	ΔMu (%)	ΔLl (%)	ΔRw (%)	ΔCw (%)	ΔN (%)
Values	-0.03	2	0.19	-4.3	-4.5	16.1	2.2	-8.2	0.0

6.5 Summary

In this chapter, a mathematical evaluation method is proposed for the interpretation of FRA results and transformer fault diagnosis. Using the proposed approach, 10 parameters of the transformer equivalent circuit are estimated from different FRA configurations, as described below:

- From EE-OC configuration, exciting inductance (L_e), core resistance (R_e), mutual inductance (M_u), and equivalent capacitance (C_{eq}) can be extracted
- EE-SC configuration gives information about windings leakage inductance (L_l) and resistance (R_w)
- The CIW scheme can be used to extract inter-winding capacitance (C_w)
- Finally, the winding turns-ratio (N) can be extracted from the IIW configuration

These extracted parameters are then used as identifiers to assess transformer winding condition. The results showed that the proposed parameter estimation strategy gives a deep insight into the transformer frequency response and aids in developing fault diagnosis methodology based on FRA results. However, to standardize this method, it is sensible to set limits to the permissible percentage change of these parameters that can serve as an objective diagnostic criterion for the abnormality of the transformer. The LFB of the EE-SC configuration is equivalent to the short circuit impedance. According to IEEE Std. C57.152-2013 [5], if the short-circuit impedance of a transformer changes by 3% or more, then the transformer is recognized as damaged. It can be appreciated by the above case studies where ΔL_l and ΔR_w for both cases are higher than 3%. Similarly, IEEE Std. C57.152-2013 defined a threshold for transformer turns-ratio that should be within 0.5% of the specified nameplate voltage for all windings. In this way, the limits for other parameters can also be defined.

For practical application of the proposed method, some points must be considered, i.e., the frequency ranges. It is important to note that frequency ranges used in this paper to extract different parameters are valid for the particular transformer windings in the example. As frequency ranges depend upon the size and rating of the transformer. Nevertheless, the frequency ranges can be identified by the characteristic frequency response for a particular transformer.

7 Artificial Intelligence for Automatic Condition Assessment

This chapter describes an artificial intelligence-based method for automatic condition assessment based on FRA measurements. The application of six types of machine learning classifiers is briefly discussed. Finally, the development of an expert system is explained for objective interpretation of FRA results.

7.1 Machine learning (ML)

Machine learning is an AI discipline that allows systems to learn and evolve behavior from examples, data, and experience. Instead of using pre-defined programming rules, a system learns from data, captures characteristics of interest, and makes intelligent decisions under uncertainty [98]. Machine learning algorithms are commonly called pattern recognition or classification algorithms. The input to the machine learning algorithm is the training data that represents the experience. The algorithms learn rules or functions to guide their performance and assign output (label) to given input (instances) [98], [99]. The detailed theoretical background of machine learning types is presented in Appendix G.

In the framework of this thesis six machine learning classification models namely Artificial neural network (ANN), decision tree (DT), random forest (RF), Naïve Bayes (NB), logistic regression (LR), and support vector machines (SVM) are studied for transformer winding condition assessment. The theoretical framework of each classifer is discussed objectively in Appendix G

To effectively evaluate and compare the performance of different classifiers, five different types of performance metrics are employed, i.e., accuracy, confusion matrix, precision, recall, and F-score. The definition of each performance metric is given in Appendix G.

7.2 Transformer Condition Assessment Algorithm (TCA)

In this section, the necessary steps for building a classifier for a given dataset are described. The first step is to analyze the data by using prior knowledge from

Chapter 3 to transform the data into a set of attributes. These attributes should represent the main characteristics of the data in a more compact manner. The performance of the different classifiers mentioned above is compared for different feature sets. Finally, based on the performance comparison, and validation with case studies, the best-performing classifier with the most appropriate feature set is selected.

7.2.1 Methodology

The necessary steps for building the Transformer Condition Assessment (TCA) algorithm for a given data set are illustrated in Figure 7-1. The process includes six major steps, i.e., Data preparation, feature generation, training of the classifier, testing, and validation with unknown data, performance analysis, and validation with selected unseen case studies.

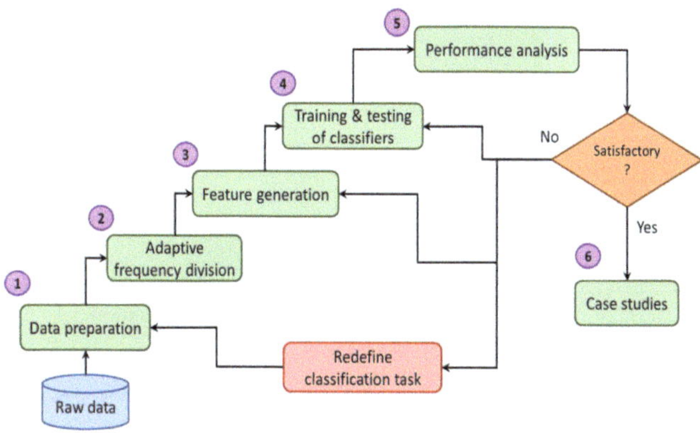

Figure 7-1: Flowchart for intelligent fault detection algorithm

If the performance of the model is not satisfactory, it can be attributed to erroneous data preparation, poor feature generation, or simply due to less diversity among deviation patterns of different classes. In such situations, the TCA algorithm recommends improving the performance by redefining the process steps.

7.2.2 Data preparation

Data pre-processing involves two steps: data labeling and noise removal. The database used in this study consists of 139 FRA results from 80 real power transformers of different designs, ratings, and different manufacturers as described in

Chapter 3. In the database, each FRA measurement belongs to a predefined state of the transformer. Six transformer conditions are identified which are healthy winding, healthy winding with core saturation, mechanical faults, shorted turns between conductors, open circuit, and repeatability issues. Each condition is labeled as a class (A to F). The description of the classes in the database is presented in Figure 7-2. Generally, the data provided may be partly accompanied by noise signals which can cause several small peaks and valleys at the lower frequencies. This noise is removed using a moving average Gaussian filter.

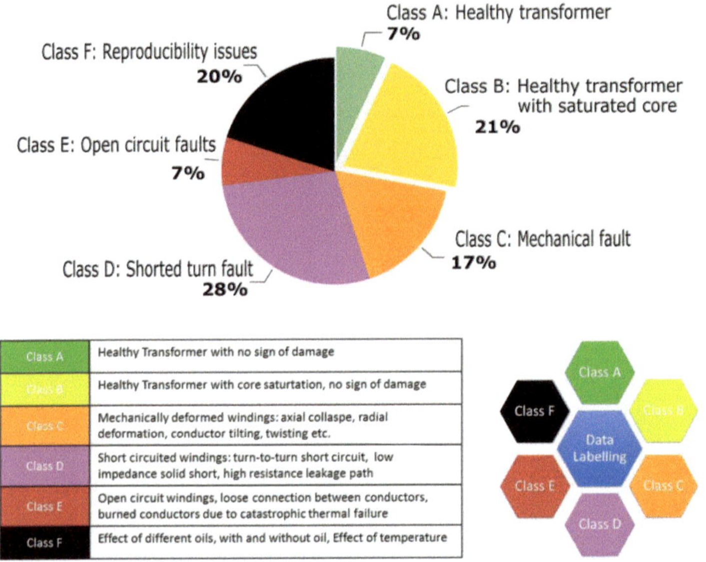

Class	Description
Class A	Healthy Transformer with no sign of damage
Class B	Healthy Transformer with core saturtation, no sign of damage
Class C	Mechanically deformed windings: axial collaspe, radial deformation, conductor tilting, twisting etc.
Class D	Short circuited windings: turn-to-turn short circuit, low impedance solid short, high resistance leakage path
Class E	Open circuit windings, loose connection between conductors, burned conductors due to catastrophic thermal failure
Class F	Effect of different oils, with and without oil, Effect of temperature

Figure 7-2: Description of classes in the database

(a) breakdown of classes in the database

(b) description of classes

The process of data de-noising is shown in Figure 7-3. The identification of the noise frequency range and the number of main anti-resonances in the low-frequency range should be determined first. Then the Gaussian filter is applied with a moving window of variable size to remove small peaks and valleys due to noise until the number of valleys in this range is equal to the defined anti-resonance points. An example of FRA signature before and after de-noising is shown in Figure 7-4. After this process, data is considered clean.

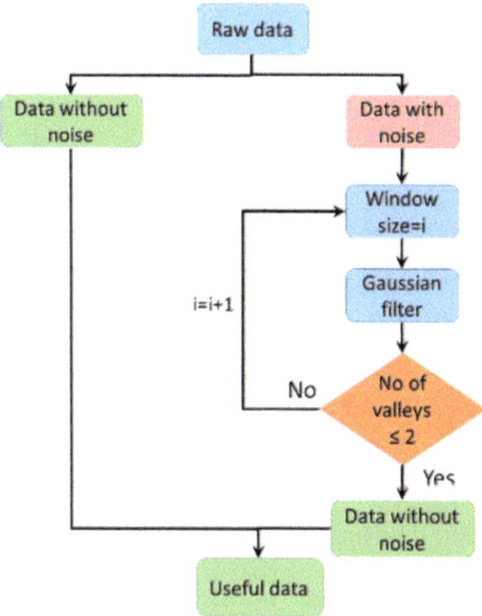

Figure 7-3: Flowchart for data de-noising algorithm

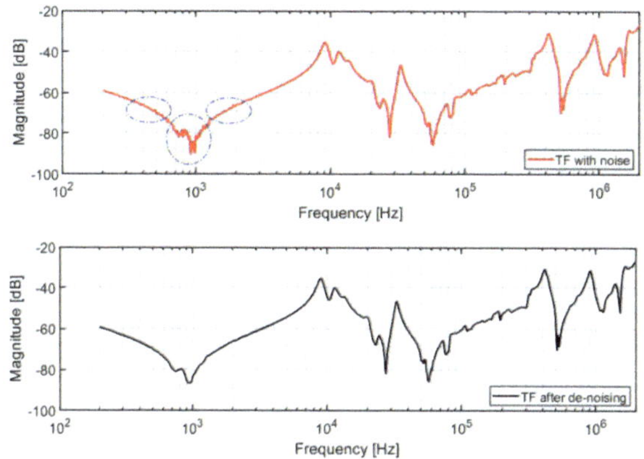

Figure 7-4: Comparison of the FRA signature before (up) and after (down) de-noising

7.2.3 Feature Generation

Feature generation also involves two steps. Firstly, the entire frequency spectrum of FRA is divided into four frequency sub-bands, i.e., two low-frequency sub-bands (LFB1 and LFB2), medium frequency sub-band (MFB), and a high-frequency sub-band (HFB). The adaptive frequency division algorithm is employed which successfully identifies low, medium, and high-frequency regions in FRA measurement depending on transformer size, rating, winding type, and connection scheme. The details of the adaptive frequency division algorithm are described in Chapter 3 [79].

Secondly, the six best-selected numerical indices from Chapter 4, i.e., LCC, SDA, SE, CCF, CSD and SD are employed to quantify the changes between reference and current FRA signatures in each frequency sub-band. Thus, each index gives 4 features from a single FRA measurement which will serve as features for the TCA algorithm. An example of frequency sub-band division and feature generation from a typical FRA measurement is presented in Figure 7-5. After frequency division and feature generation, it is possible to generate deviation patterns for the given six conditions of the transformer and these deviation patterns will serve as features for the TCA algorithm.

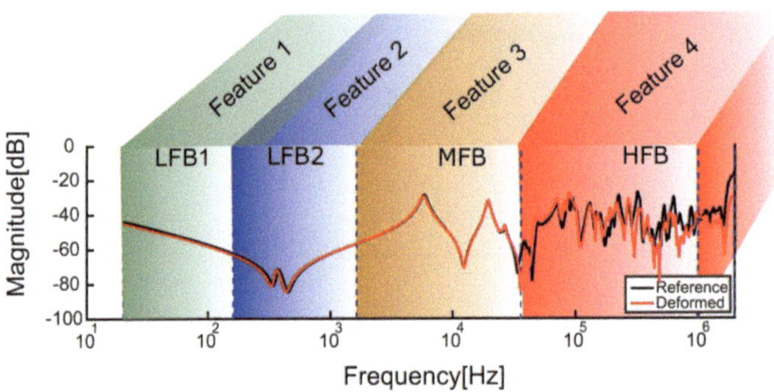

Figure 7-5: Graphical representation of frequency sub-band division and feature generation in a typical FRA measurement

7.2.4 Training and testing of classifiers

For the classification of considered faults (Class A to F), all the classifiers are trained on the same data. The input provided is a feature matrix (133 × 24) and the label matrix (133 × 1). While six cases are reserved for case studies. The label matrix is the class of each case, such as class A to F. The classifiers learn from the training data and predict output (the class of the test data). The performance comparison of different classifiers has the following goals:

- Estimate the general accuracy and predictive performance of the model on unseen data.
- Identification of feature set with the highest performance scores.

- Selection of the best-performing classifier for the problem. The model evaluation and selection technique used in this work is the K-fold cross-validation method.

K-fold cross-validation: To ensure that every observation from the data set appears in the train and test sets, the K-fold cross-validation technique is used. In this method, data is divided into K subsets (folds), and then the model is trained using K-1 folds and tested on the Kth fold. This procedure is repeated K times so that every fold is tested once as illustrated in Figure 7-6. This procedure results in K different models fitted on different yet partly overlapping training sets and tested on non-overlapping test sets.

Eventually, the cross-validation performance is computed as the arithmetic mean over the K performance estimates. The choice of K is arbitrary and there is a bias-variance tradeoff associated with it. Usually, 5 or 10 is an ideal choice. Taking into account the small data set, 5-fold stratified cross-validation is used for the combined feature set. StratifiedKfold returns stratified folds, i.e., each fold contains approximately the same percentage of samples of each target class.

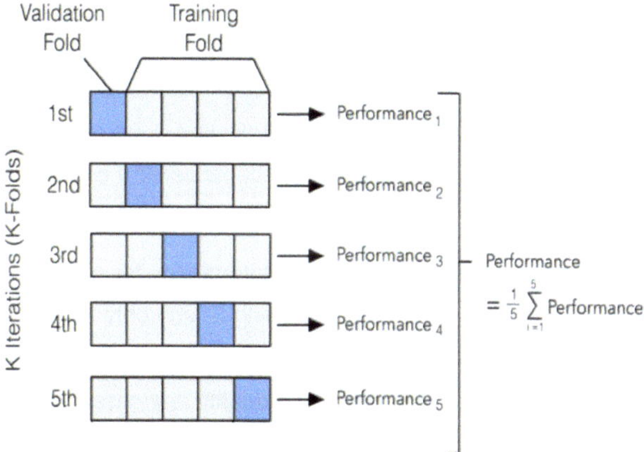

Figure 7-6: Representation of 5-fold cross-validation technique

7.2.5 Performance Analysis

At this stage, the performance of the algorithm is evaluated via performance evaluation metrics. Mainly, five performance evaluation metrics, i.e., accuracy, precision, recall, F-score, and confusion matrix are employed in this research, which gives general and detailed (per class) performance of the algorithm. The predictive performance of different classification models for given data is summarized using confusion matrices as illustrated in Figure 7-7. These confusion matrices are obtained using the 5-fold cross-validation method in Python. The diagonal cells show the number of correctly classified cases for each class, while off-diagonal cells represent the number of misclassified cases. The performance measure used here is the mean accuracy of 5-fold, which is calculated as the ratio of correctly classified cases to the total number of cases in the dataset. The performance analysis of each classifier is briefly explained in the following:

Decision Tree: DT shows good performance in predicting classes A, C, D, and F as for each class it misclassified 2 cases only. All cases of class B are correctly identified with 100% accuracy and only 1 case for class E is misclassified leading to excellent performance for these classes.

Random Forest: The performance of RF in predicting classes B, C, and D is excellent as no case of class B, 2 cases of class C, and only 1 case of class D are misclassified. While predicting performance for classes A and E is good with 3 misclassified cases for each class. The only class with acceptable performance is class F as 4 cases are misidentified by the RF as class C and D.

Naive Bayes (NB): NB shows excellent performance for classes A, B, D, E, and F, as only 1 case of class F and 2 cases of classes A, B, and D are misclassified. However, the predictive performance for class C is acceptable, as 4 cases of class C are misclassified.

Support Vector Machine: SVM shows excellent performance for classes B, C, D, and E. It correctly identified all cases of class B and E, and only 2 cases of C and D are misidentified. For class F, it shows good performance as only 3 cases are not predicted correctly. However, SVM fails to identify any case for class A, as the majority of cases of class A (14 cases) are misclassified as class C and 2 cases are misclassified as class B. Hence, the accuracy of SVM in predicting class A is 0%. The possible reason behind this is that the FRA traces of a slight mechanical change for a power transformer are similar to the FRA traces of a normal transformer. For SVM, it becomes difficult to draw a hyper-plane between classes A and C as the features of these classes are very similar and linearly inseparable.

Logistic Regression: LR shows similar trends as observed with SVM. The predictive performance of LR is excellent for classes B, C, D, and E. It misclassified 1 case of class C and 2 cases of class D while all cases of class B and E are predicted with 100% accuracy. For class F, it misidentified 4 cases leading to an acceptable performance for this class. In class A, only one case is correctly classified while 15 cases are misclassified as class C. This misclassification can be attributed to the linear nature of LR, as features of marginal mechanical faults (class C) and healthy windings (class A) are very similar.

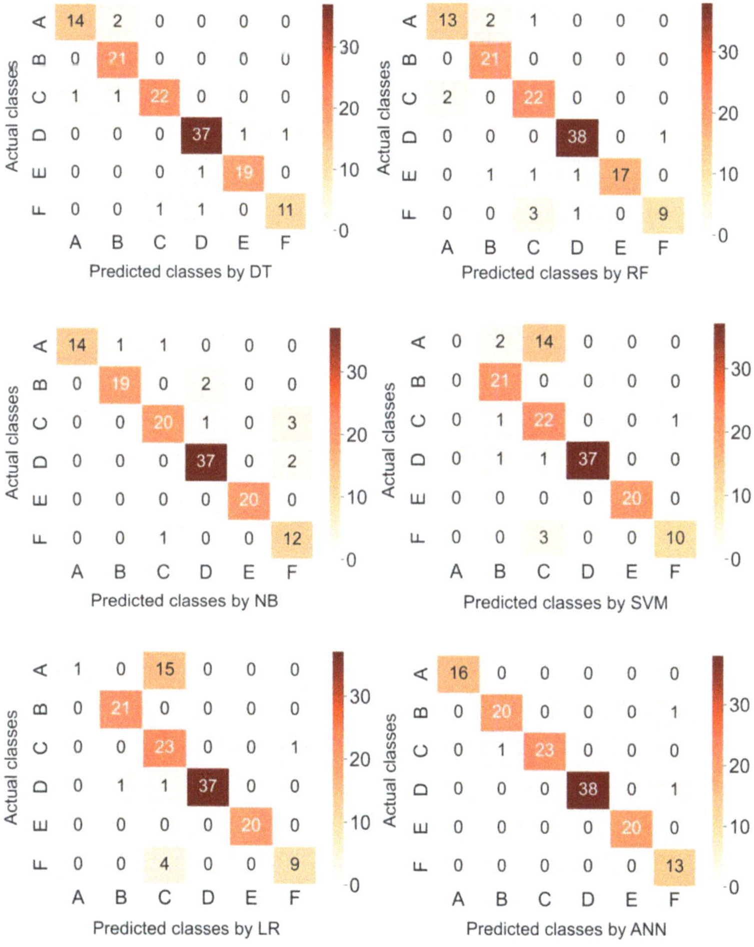

Figure 7-7: Confusion matrices for different machine learning classifiers

Artificial neural network: The performance of ANN in predicting classes A, E, and F is excellent as all the cases are correctly identified with 100% accuracy in these classes. While, predicting performance for classes B, C, and D is very good as there is only 1 misclassified case for each class.

Figure 7-8 shows the general performance of the classifiers in 5-fold cross-validation for a combined feature set in which all indices (CCF, LCC, CSD, SD, SDA, and SE) are combined to make one feature set. The performance is expressed by the mean accuracy of 5-fold and standard deviation (std). The highest accuracy is obtained with ANN (96%), followed by DT (92%), NB (92%), and RF (90%). While SVM (82%), and LR (82%) show the least accuracy. The deviation from the mean value is lowest in RF (2%) and ANN (3%) which means that RF and ANN show fewer variations than any other classifier while predicting class labels in different folds.

Even though the general classification performance of all of the classifiers is good, a detailed analysis of classification performance per class is done. Under the premise of imbalanced class distribution, a classifier can exhibit good performance for specific classes (usually majority classes) which does not necessarily imply that the classification of all classes is good. For this purpose, precision, recall, and F-score metrics (PRF) are used. A comparison of PRF calculated for each class against different classifiers is presented in Figure 7-9, Figure 7-10, and Figure 7-11, respectively.

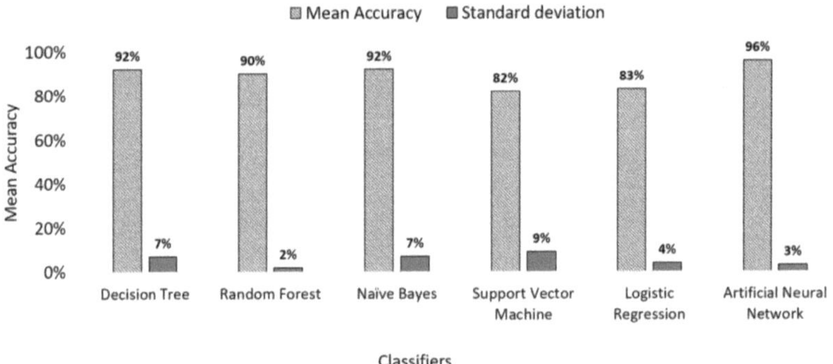

Figure 7-8: Mean accuracy and standard deviation of classifiers in 5-fold cross-validation for combined feature set

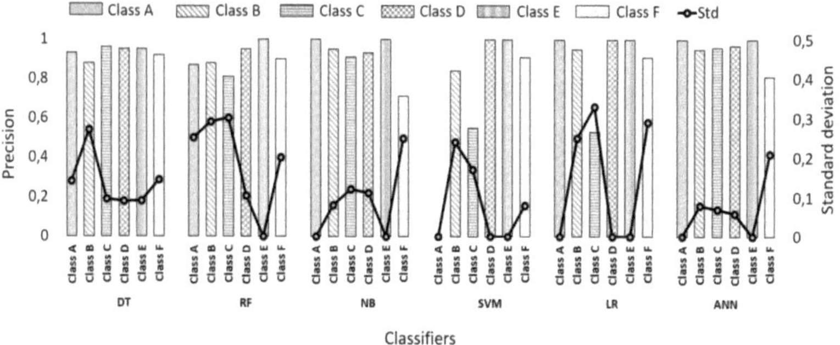

Figure 7-9: Mean precision and standard deviation of classifiers in 5-fold cross-validation for combined feature set

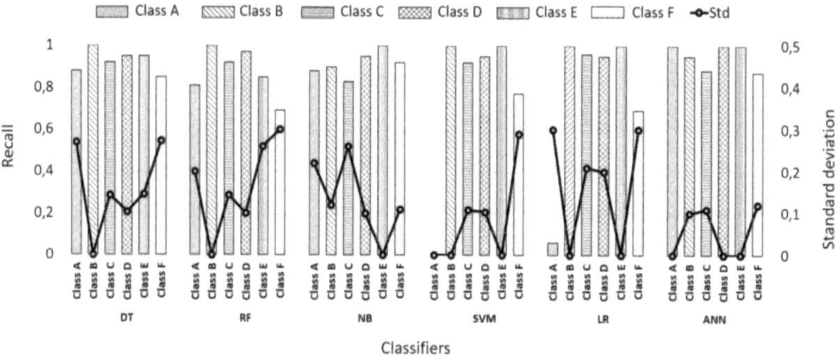

Figure 7-10: Mean recall and standard deviation of classifiers in 5-fold cross-validation for combined feature set

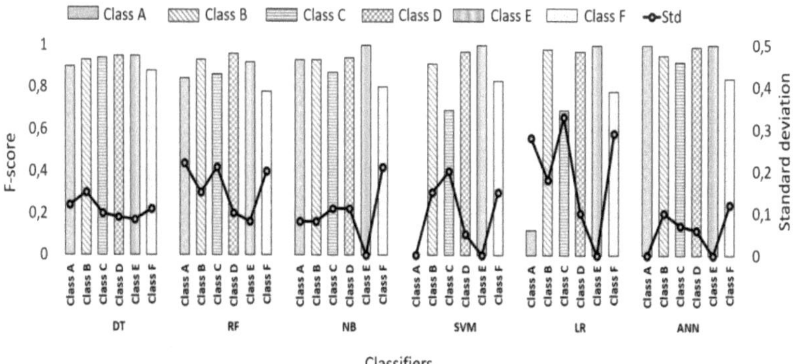

Figure 7-11: Mean F-score and standard deviation of classifiers in 5-fold cross-validation for the combined feature set

The precision chart gives the information of all the transformers that are predicted as healthy, and how many are actually healthy. It can be appreciated from Figure 7-9 that ANN exhibits excellent values of precision as the average precision is above 95% with the lowest precision for class F. It also possesses the least standard deviation values. The average precision for the DT classifier is above 90% with the lowest value for class B. While RF and NB show the average precision above 85%. LR and SVM show the lowest values of precision in different classes. A similar trend can be observed in Recall and F-score results which confirms the consistency and reliability of the classifiers.

7.2.6 Case studies

7.2.6.1 Case 1: Axial collapse after clamping failure

In case 1, the unit is a 3-phase, 240 MVA, 400/132 kV autotransformer [13]. The unit was switched out of service for investigation after a Buchholz alarm. FRA measurements on common winding before and after the fault are shown in Figure 7-12. The visual analysis of the FRA results shows some deviations and shifts of resonances in the high-frequency sub-bands. After strip-down irreparable damage such as axial collapse to the LV winding was found as shown in Figure 7-12. In order to verify the performance of the ML classifiers in diagnosing this case. The case is further tested with the six classifiers DT, RF, NB, LR, SVM, and ANN. These classifiers are trained with a combined set of indices (CSD, SD, CCF, LCC, and SE). The performance metrics are shown in Table 7-1. It can be seen that all the classifiers have successfully diagnosed this case as a mechanical fault.

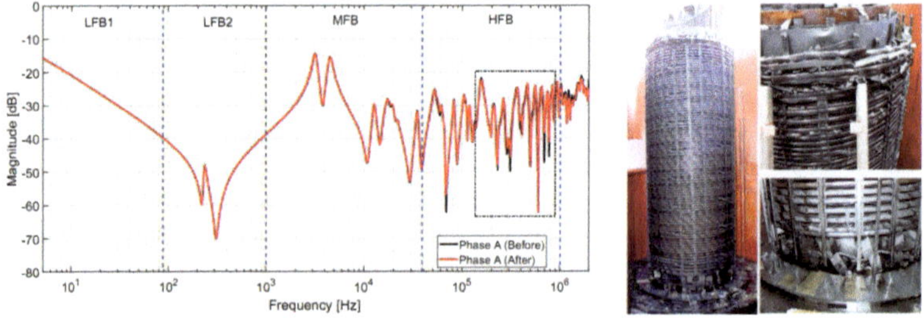

Figure 7-12: LV winding TF before and after fault (left) and inspection of fault after dismantling (right)

Table 7-1: Performance analysis of classifiers: Case 1

Classifiers	Actual class	Predicted class	Comment
Decision Tree	C	C	Pass
Random Forest	C	C	Pass
Naive Bayes	C	C	Pass
Support Vector Machine	C	C	Pass
Logistic Regression	C	C	Pass
Artificial Neural Network	C	C	Pass
C: Mechanically deformed winding			

7.2.6.2 Case 2: Shorted turn failure

In case 2, the unit is a 3-phase, 60 MVA, 105/6.6/22 kV transformer. The unit was tested after the lightning strike. Figure 7-13 shows the HV EE-OC measurements before and after the fault. The response of the short-circuited winding is deviated in the low and medium frequency bands. After dismantling the transformer, it was found that phase-U turns shorted due to a lightning surge. This case is further tested with six ML classifiers. The performance metrics are shown in Table 7-2. It can be appreciated that all the classifiers have diagnosed the correct class.

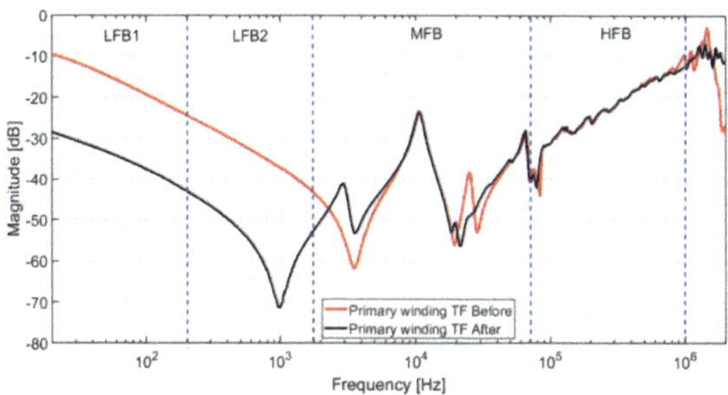

Figure 7-13: HV winding TF before and after fault

Table 7-2: Performance analysis of classifiers: Case 2

Classifiers	Actual class	Predicted class	Comment
Decision Tree	D	D	Pass
Random Forest	D	D	Pass
Naive Bayes	D	D	Pass
Support Vector Machine	D	D	Pass
Logistic Regression	D	D	Pass
Artificial Neural Network	D	D	Pass
D: Shorted turn fault			

7.2.6.3 Case 3 and 4: Marginal mechanical faults (Axial and radial deformation) in HV winding

In the 3rd and 4th case studies, the FRA traces of a 27 MVA, 154/10.5 kV transformer are employed [86]. In case 3, the V-phase HV winding is radially deformed to discuss the diagnosis of radial deformation by FRA. The open-circuit FRA traces and damaged winding is shown in Figure 7-14. In case 4, HV winding is axially displacement by 0.6% of its height. The open-circuit FRA traces and damaged winding can be seen in Figure 7-15.

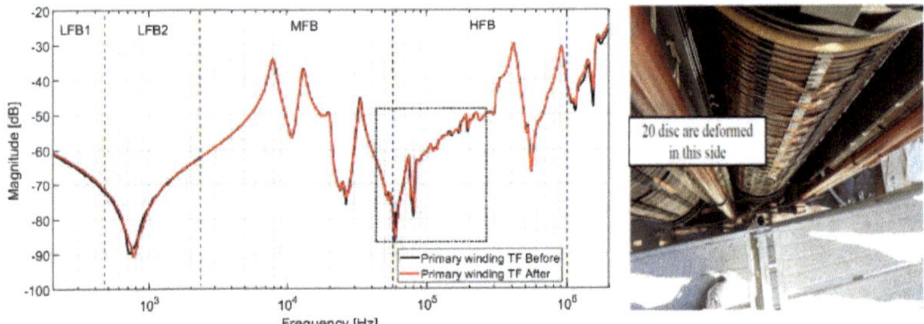

Figure 7-14: HV winding TF before and after fault (left) and deformed winding (right)

Table 7-3: Performance analysis of classifiers: Case 3

Classifiers	Actual class	Predicted class	Comment
Decision Tree	C	C	Pass
Random Forest	C	C	Pass
Naive Bayes	C	C	Pass
Support Vector Machine	C	F	Fail
Logistic Regression	C	F	Fail
Artificial Neural Network	C	C	Pass

C: Mechanically deformed winding, F: Reproducibility issues

From Figure 7-14 and Figure 7-15, it can be seen that marginal faults are difficult to interpret manually, as the deviations between FRA curves are very small. However, with the application of the ML classifiers, it is possible to detect such slight deviations between FRA curves. The performance metrics are shown in Table 7-3 and Table 7-4, respectively. It can be appreciated that DT, RF, NB, and ANN have correctly diagnosed radial deformation as a mechanical failure. However, LR and SVM failed to detect the radial deformation, indicating class F (reproducibility issues). A very similar behavior was observed in case 4, where DT, RF, and ANN have correctly classified the case. However, NB, LR, and SVM failed to detect the axial displacement, indicating class F (reproducibility issues).

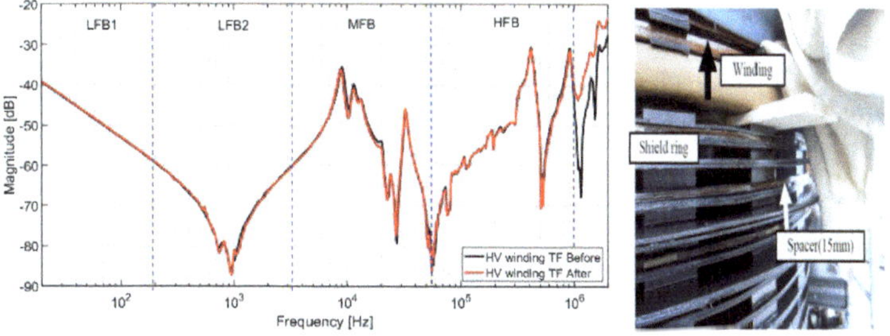

Figure 7-15: HV winding TF before and after fault (left) and deformed winding (right)

Table 7-4: Performance analysis of classifiers: Case 4

Classifiers	Actual class	Predicted class	Comment
Decision Tree	C	C	Pass
Random Forest	C	C	Pass
Naive Bayes	C	F	Fail
Support Vector Machine	C	F	Fail
Logistic Regression	C	F	Fail
Artificial Neural Network	C	C	Pass

C: Mechanically deformed winding, F: Reproducibility issues

7.2.6.4 Case 5: Healthy transformer

In case 5, the unit is a 3-phase, 47 MVA, 120/26.4 kV transformer. This case belongs to the class of normal transformers which have successfully passed the short-circuit test. The FRA traces are measured before and after the short-circuit test event. After inspection, no winding deformations were found in this transformer. The FRA curves of HV winding are presented in Figure 7-16. It can be seen that the FRA curves perfectly lie on each other, indicating no deviation except in the low-frequency regions where the deviations are due to a different core magnetization. The case is further tested with the six ML classifiers. The results are presented in Table 7-5. It can be appreciated that DT, RF, NB, and ANN have diagnosed this case as a healthy transformer with core saturation. However, SVM and LR failed to correctly identify the fault, indicating class C (mechanically deformed winding).

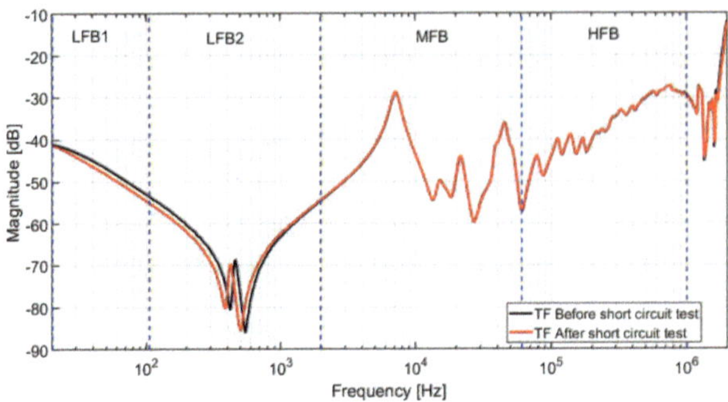

Figure 7-16: HV winding TF before and after short circuit event

Table 7-5: Performance analysis of classifiers: Case 5

Classifiers	Actual class	Predicted class	Comment
Decision Tree	B	B	Pass
Random Forest	B	B	Pass
Naive Bayes	B	B	Pass
Support Vector Machine	B	C	Fail
Logistic Regression	B	C	Fail
Artificial Neural Network	B	B	Pass

C: Mechanically deformed winding, B: Healthy Transformer with saturated core

7.2.6.5 Case 6: Open circuit fault

In case 6, the unit is a 3-phase, 34 MVA, 237/5.65 kV transformer that was switched out of service for investigation after a Buchholz alarm [13]. The FRA traces before and after the fault are shown in Figure 7-17. At low frequency, an increase in the attenuation accompanied by a vertical shift downwards can be observed which indicates that the open circuit does not completely break the electrical connection of the winding. At high frequencies, the peaks do not show frequency shifts, but the vertical shifts are a clear indication of the change in resistance of the winding. After dismantling it was confirmed an open circuit fault as shown in Figure 7-17.

The case is further tested with the six ML classifiers and results are presented in Table 7-6. It can be appreciated that all the classifiers have correctly identified the type of fault, indicating class E (open circuit winding).

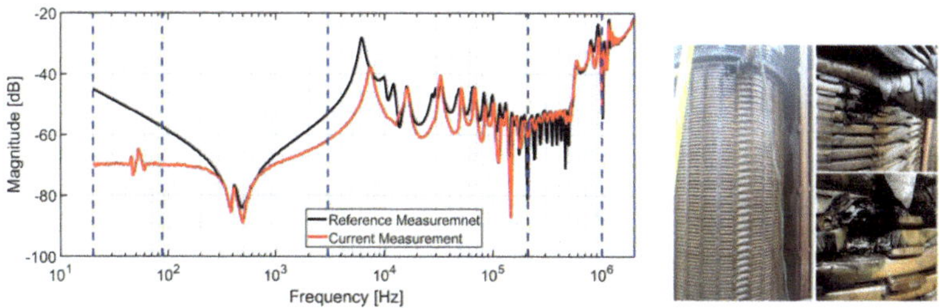

Figure 7-17: HV winding TF before and after fault (left) and deformed winding (right)

Table 7-6: Performance analysis of classifiers: Case 6

Classifiers	Actual class	Predicted class	Comment
Decision Tree	E	E	Pass
Random Forest	E	E	Pass
Naive Bayes	E	E	Pass
Support Vector Machine	E	E	Pass
Logistic Regression	E	E	Pass
Artificial Neural Network	E	E	Pass

E: Open circuit winding

7.3 Summary

In this chapter, five machine learning models, namely, Decision Trees, Random Forest, Naive Bayes, Support Vector Machine, Logistic Regression, and Artificial Neural Network are studied for a reliable and automatic assessment of transformers FRA results. Six states of the transformer are considered within the database of 139 FRA measurements from 80 power transformers of different designs, sizes, and different manufacturers. The classification capabilities of these classifiers are compared based on accuracy in the 5-fold cross-validation method. The combined set of indicators is used to train the classifiers.

It was found that the ANN, DT, RF, and NB exhibit good performance, as 90% of the cases are correctly identified by these classifiers. Among these classifiers, ANN and RF possess high accuracy and lowest standard deviation values. Thus, these two classifiers stand out from the rest of the classifiers. The performance was also validated using selected case studies from real power transformers with known diagnoses. ANN and RF have correctly classified all cases in the case study, indicating these classifiers have outplayed the others. It is important to mention that, the performance of these classifiers can be further optimized by hyperparameter tuning, by removing unnecessary feature sets in the database. Moreover, it is sensible to develop an ensemble ML classifier model based on the best performance classifiers. Results obtained in this chapter give evidence that the machine learning methods are able to precisely assess the transformer winding condition and also identify the fault type with good accuracy without much human intervention.

8 Conclusion and Future Work

This thesis presents a number of novel intelligent techniques and approaches to deal with power transformer winding distortion and deformation assessment problems based on frequency response analysis. Based on real case studies, it provides a set of guidelines and criteria for the interpretation of transformer frequency response measurements. The proposed solutions can provide an improved understanding of the health condition of power transformers that will facilitate the process of making decisions about their maintenance and replacement. The summary of the contribution of this research is given in the following:

Data collection of use cases

The FRA technique is a diagnostic method that compares a reference trace (fingerprint) and a newly measured FRA trace for the determination of the mechanical condition of the transformer. In order to test different interpretation techniques, a data set of FRA with reference traces and traces of known mechanical defects is necessary. In the framework of this research, the data is collected from different utilities, and diagnosis companies in collaboration with partner universities, institutes, and CIGRE WG A2.53. Additionally, the FRA measurements, where winding deformations are simulated in different transformers were also used.

Development of adaptive frequency division algorithms

As a first attempt towards automation of the assessment of FRA results, the necessity of dividing the frequency spectrum into sub-bands has been recognized as an approach that enhances the reliability of the assessment. The frequency ranges in FRA are specific to each winding and hence, it is very meaningful to develop an algorithm for identification of transformer-specific frequency sub-bands. In this research, sub-band identification algorithms are presented and formulated as the basis for a reliable and automatic assessment of FRA results. The suitability of the algorithm is verified with real case studies. The algorithm satisfactorily identifies low, medium, and high-frequency sub-bands in different FRA signatures considering the transformer winding type and voltage range.

Investigation on the effects of different failure modes in FRA

The characterization of the effects of various failure modes and reproducibility problems in different frequency sub-bands is crucial for the reliable assessment of FRA results because based on these characterization non-expert users can also have a better understanding of the deviations between two FRA traces. In this research, the effects of electrical, and mechanical failures and reproducibility problems are discussed from real case studies. Moreover, the most affected frequency sub-bands under the influence of different faults are also characterized.

Development and evaluation of numerical indicators

In the framework of this research, various indices are collected on a single platform, compared and categorized with some newly proposed indices based on their characteristics, to announce more appropriate indices as the standard ones. The analysis of the indices was performed with real case studies. This contribution recommends the four indices LCC, SDA, SE, and CCF as the most appropriate ones that can satisfactorily define the extent of change between two TFs.

Development of a winding assessment factor

To overcome the drawbacks of conventional numerical indicators, a winding assessment factor is introduced which is based on the sliding window method. The method is tested with different case studies of real transformers with normal and deformed windings. Based on the presented case studies, a tentative criterion for transformer fault detection is proposed (MSDD<-5). Thus, with these investigations, an objective criterion has been established for power transformer fault detection by frequency response analysis which was a key task for an effective assessment tool.

Modelling of transformer frequency response by FEM software

In this contribution, the physical geometry of a three-phase transformer is simulated using 3D finite integration analysis to emulate the real transformer operation. The novelty of this model is that FRA traces are directly obtained from the 3D model of windings without estimating and solving lumped parameter circuit models. At first, the method is validated with a simple experimental setup. Afterward, different mechanical and electrical faults are simulated, and their effects on FRA are discussed in different FRA test configurations objectively. Such a model is useful to study the impact of different faults on FRA, and to generate a large database that can be used to train the machine learning algorithms.

Transformer equivalent circuit parameter estimation

A simple mathematical approach is introduced to estimate the main parameters of the equivalent circuit of a power transformer from different types of FRA measurements. For this purpose, a sensitivity study is performed using the HF FRA simulation model. Mainly, 10 parameters of the transformer equivalent circuit are estimated from different FRA configurations. The percentage change of these parameters can serve as an objective-interpretation methodology for diagnosis of transformer electrical and mechanical faults.

Development of intelligent classification methods based on AI

In this research, an expert system is developed based on a combined approach using numerical indices, and a supervised machine learning technique to develop a method for automatic condition assessment of transformers winding using FRA results. As a first attempt towards automatic assessment of FRA results, this algorithm is implemented in a graphical user interface to develop an FRA app. For this purpose, the performance of six widely used machine learning algorithms, i.e., Artificial Neural Network (ANN), Decision Tree (DT), Random Forest (RF), Naive Bayes (NB), Support Vector Machine (SVM), and Logistic Regression (LR) is studied for transformer winding condition assessment by testing them with real case studies. The highest accuracy was obtained with ANN (96%), followed by DT (92%), NB (92%), and RF (90%). Results obtained in this research give evidence that the proposed expert system is able to precisely assess the transformer winding condition and also identify the fault type with good accuracy without much human intervention.

8.1 Recommendations for Future Research

More case studies from the field are necessary to settle down the proposed criterion of the transformer winding assessment factor. Moreover, the proposed criterion is only valid for time-based FRA comparisons. Hence, it is sensible to develop different criteria for type-based, and construction-based FRA comparisons. Such a study demands a large amount of FRA data from different transformers having various faults including the datasets of healthy transformers.

The proposed HF model can be used to generate a large database that covers all the possible winding faults. This database can be used to study and generate different deviation patterns associated to different faults using proposed indicators. These deviation patterns can be served as features to train and develop an intelligent fault detection and classification algorithm, i.e., artificial neural network. Additionally, the application of the HF model should be extended to study the effects of different factors (moisture effect, temperature effect, aging, etc.) on the frequency response. In addition, the proposed model can also be employed for studying the transient analysis of power transformers, and calculation of winding overvoltages under different fault conditions.

In this research, a method is presented to extract the key parameters of the transformer equivalent circuit. However, due to limited datasets from the field, the method is only tested on small transformer models where faults are simulated. Thus, it is sensible to test this method on real case studies and it is also possible to set a threshold on percentage changes to different parameters. That can serve as a meaningful transformer winding assessment criterion.

In the framework of this research, mainly three different types of indicators are presented to detect variations between two FRA measurements, i.e., numerical indicators, winding assessment factor, and transformer equivalent circuit parameters. Each type of indicator has its own advantages and disadvantages. Thus, it is possible to combine all three types of indicators. The deviation features generated by these indicators can increase the learning rate of the machine learning classifiers. Thus, the accuracy of the assessment can be increased.

As a first attempt to develop an automatic assessment tool a GUI is developed which is based on a single classifier, i.e., ANN, as it showed the best performance. However, it is sensible to develop an artificial intelligence tool based on the multiple machine learning classifiers which are based on ensemble machine learning techniques. In this way, the accuracy and reliability of the assessment can be further enhanced.

Finally, an asset management tool is the need of the hour that cannot only store and manage all the FRA datasets but it should also serve as an automatic assessment tool for transformer FRA results. The FRA app can be further extended to meet these requirements. Such an app can be announced as an online FRA tool that will be available to the public. It will help the transformer community to upload and interpret FRA results. Moreover, it will also help to collect the FRA database worldwide.

9 Bibliography

[1] CIGRE Working Group A2.18, *Life management techniques for power transformer.* CIGRE Brochure 227; CIGRÉ Paris, 2003.

[2] D. Kopejtkova, et al., 'Strategy for condition based maintenance', *CIGRÉ Paris Sess. 1996*, pp. 23–105, 1996.

[3] A. J. McGrail, et al., 'Data Mining Techniques to Assess the Condition of High Voltage Electrical Plant', in *CIGRÉ Session 2002*, Paris, 2002, pp. 15–107.

[4] S. Tenbohlen, S. Coenen, M. Djamali, A. Müller, M. Samimi, and M. Siegel, 'Diagnostic Measurements for Power Transformers', *Energies*, vol. 9, no. 5, p. 347, May 2016, doi: 10.3390/en9050347.

[5] IEEE Std. C57.149-2012, *IEEE Guide for the Application and Interpretation of Frequency Response Analysis for Oil-Immersed Transformers*, 2012. doi: 10.1109/IEEESTD.2013.6475950.

[6] X. Zhao, C. Yao, C. Zhang, and A. Abu-Siada, 'Toward reliable interpretation of power transformer sweep frequency impedance signatures: experimental analysis', *IEEE Electr. Insul. Mag.*, vol. 34, no. 2, pp. 40–51, Mar. 2018, doi: 10.1109/MEI.2018.8300443.

[7] CIGRE Working Group A2.26, *Mechanical condition assessment of transformer windings using frequency response analysis (FRA).* CIGRE Brochure 342;, 2008.

[8] E. Rahimpour, J. Christian, K. Feser, and H. Mohseni, 'Transfer function method to diagnose axial displacement and radial deformation of transformer windings', *IEEE Trans. Power Deliv.*, vol. 18, no. 2, pp. 493–505, Apr. 2003, doi: 10.1109/TPWRD.2003.809692.

[9] R. Wimmer, *Die Ermittlung der Übertragungsfunktion von Großtransformatoren mittels On- und Offline-Messungen*, 1. Aufl. in Schriftenreihe des Instituts für Energieübertragung und Hochspannungstechnik, no. 2. Göttingen: Sierke, 2010.

[10] M. H. Samimi, A. A. Shayegani Akmal, H. Mohseni, and S. Tenbohlen, 'Detection of transformer mechanical deformations by comparing different FRA connections', *Int. J. Electr. Power Energy Syst.*, vol. 86, pp. 53–60, Mar. 2017, doi: 10.1016/j.ijepes.2016.09.007.

[11] IEC60076-18, *Measurement of frequency response*, 2012.

[12] P. Picher, S. Tenbohlen, M. Lachman, A. Scardazzi, and P. Patel, 'Current state of transformer FRA interpretation', *Procedia Eng.*, vol. 202, pp. 3–12, 2017, doi: 10.1016/j.proeng.2017.09.689.

[13] CIGRE Working Group A2.53, *Advances in the interpretation of transformer Frequency Response Analysis (FRA).* CIGRE Brochure 812; CIGRÉ Paris, 2020.

[14] N. Hashemnia, A. Abu-Siada, and S. Islam, 'Improved power transformer winding fault detection using FRA diagnostics – part 1: axial displacement

simulation', *IEEE Trans. Dielectr. Electr. Insul.*, vol. 22, no. 1, pp. 556–563, Feb. 2015, doi: 10.1109/TDEI.2014.004591.

[15] N. Abeywickrama, Y. V. Serdyuk, and S. M. Gubanski, 'High-Frequency Modeling of Power Transformers for Use in Frequency Response Analysis (FRA)', *IEEE Trans. Power Deliv.*, vol. 23, no. 4, pp. 2042–2049, Oct. 2008, doi: 10.1109/TPWRD.2008.917896.

[16] H. Zhang, S. Wang, D. Yuan, and X. Tao, 'Double-Ladder Circuit Model of Transformer Winding for Frequency Response Analysis Considering Frequency-Dependent Losses', *IEEE Trans. Magn.*, vol. 51, no. 11, pp. 1–4, Nov. 2015, doi: 10.1109/TMAG.2015.2442831.

[17] E. Rahimpour, M. Jabbari, and S. Tenbohlen, 'Mathematical Comparison Methods to Assess Transfer Functions of Transformers to Detect Different Types of Mechanical Faults', *IEEE Trans. Power Deliv.*, vol. 25, no. 4, pp. 2544–2555, Oct. 2010, doi: 10.1109/TPWRD.2010.2054840.

[18] M. M. H. Samimi and S. Tenbohlen, 'The Numerical Indices Proposed For the Interpretation of the FRA Results: A Review', VDE-Hochspannungstechnik , Berlin 2016.

[19] M. Bigdeli, M. Vakilian, and E. Rahimpour, 'Transformer winding faults classification based on transfer function analysis by support vector machine', *IET Electr. Power Appl.*, vol. 6, no. 5, p. 268, 2012, doi: 10.1049/iet-epa.2011.0232.

[20] O. Aljohani and A. Abu-Siada, 'Application of DIP to Detect Power Transformers Axial Displacement and Disk Space Variation Using FRA Polar Plot Signature', *IEEE Trans. Ind. Inform.*, vol. 13, no. 4, pp. 1794–1805, Aug. 2017, doi: 10.1109/TII.2016.2626779.

[21] M. Bigdeli, D. Azizian, H. Bakhshi, and E. Rahimpour, 'Identification of transient model parameters of transformer using genetic algorithm', in *2010 International Conference on Power System Technology*, Zhejiang, Zhejiang, China: IEEE, Oct. 2010, pp. 1–6. doi: 10.1109/POWERCON.2010.5666462.

[22] A. Holdyk, B. Gustavsen, I. Arana, and J. Holboell, 'Wideband Modeling of Power Transformers Using Commercial sFRA Equipment', *IEEE Trans. Power Deliv.*, vol. 29, no. 3, pp. 1446–1453, Jun. 2014, doi: 10.1109/TPWRD.2014.2303174.

[23] D. Filipović-Grčić, et al., 'High-frequency model of the power transformer based on frequency-response measurement', *IEEE Transactions on Power Delivery*, vol. VOL. 30, NO. 1, Feb. 2015.

[24] V. S. Larin, et al., 'Application of natural frequencies deviations patterns and high-frequency white-box transformer models for FRA', Cigre Paris 2018, pp. A2-209.

[25] J. L. V. Contreras, M. A. Sanz-Bobi, S. Banaszak, and M. Koch, 'Application of machine learning techniques for automatic assessment of FRA measurements', ISH Germany 2011.

[26] K. R. Gandhi and K. P. Badgujar, 'Artificial neural network based identification of deviation in frequency response of power transformer windings', in

2014 Annual International Conference on Emerging Research Areas: Magnetics, Machines and Drives (AICERA/iCMMD), 2014, pp. 1–8. doi: 10.1109/AICERA.2014.6908217.

[27] Y. Luo *et al.*, 'Recognition technology of winding deformation based on principal components of transfer function characteristics and artificial neural network', *IEEE Trans. Dielectr. Electr. Insul.*, vol. 24, no. 6, pp. 3922–3932, 2017, doi: 10.1109/TDEI.2017.006655.

[28] X. Mao, Z. Wang, P. Jarman, and A. Fieldsend-Roxborough, 'Winding Type Recognition through Supervised Machine Learning using Frequency Response Analysis (FRA) Data', in *2019 2nd International Conference on Electrical Materials and Power Equipment (ICEMPE)*, 2019, pp. 588–591. doi: 10.1109/ICEMPE.2019.8727354.

[29] J. Liu, Z. Zhao, C. Tang, C. Yao, C. Li, and S. Islam, 'Classifying Transformer Winding Deformation Fault Types and Degrees Using FRA Based on Support Vector Machine', *IEEE Access*, vol. 7, pp. 112494–112504, 2019, doi: 10.1109/ACCESS.2019.2932497.

[30] Z. Zhao, C. Yao, C. Tang, C. Li, F. Yan, and S. Islam, 'Diagnosing Transformer Winding Deformation Faults Based on the Analysis of Binary Image Obtained From FRA Signature', *IEEE Access*, vol. 7, pp. 40463–40474, 2019, doi: 10.1109/ACCESS.2019.2907648.

[31] L. Duan, J. Hu, G. Zhao, K. Chen, S. X. Wang, and J. He, 'Method of interturn fault detection for next-generation smart transformers based on deep learning algorithm', *High Volt.*, vol. 4, no. 4, pp. 282–291, Dec. 2019, doi: 10.1049/hve.2019.0067.

[32] A. J. Ghanizadeh and G. B. Gharehpetian, 'ANN and cross-correlation based features for discrimination between electrical and mechanical defects and their localization in transformer winding', *IEEE Trans. Dielectr. Electr. Insul.*, vol. 21, no. 5, pp. 2374–2382, 2014, doi: 10.1109/TDEI.2014.004364.

[33] *FRANEO 800, SFRA-Frequenzgangsanalysator brochure*. [Online]. Available: https://www.omicronenergy.com/de/produkte/franeo-800/

[34] CIGRE Working Group A2.37, *Transformer reliability survey*. CIGRE Brochure 642; CIGRÉ Paris, 2015.

[35] IEEE Std C57.125™-2015, 'IEEE Guide for Failure Investigation, Documentation, Analysis, and Reporting for Power Transformers and Shunt Reactors', IEEE. doi: 10.1109/IEEESTD.2015.7363719.

[36] E. P. Dick and C. C. Erven, 'Transformer Diagnostic Testing by Frequuency Response Analysis', *IEEE Trans. Power Appar. Syst.*, vol. PAS-97, no. 6, pp. 2144–2153, 1978, doi: 10.1109/TPAS.1978.354718.

[37] J. C. Gonzales Arispe and E. E. Mombello, 'Detection of Failures Within Transformers by FRA Using Multiresolution Decomposition', *IEEE Trans. Power Deliv.*, vol. 29, no. 3, pp. 1127–1137, Jun. 2014, doi: 10.1109/TPWRD.2014.2306674.

[38] R. Wimmer, S. Tenbohlen, K. Feser, A. Kraetge, M. Krüger, and J. Christian, 'Development of Algorithms to Assess the FRA'.

[39] M. Heindl, S. Tenbohlen, A. Kraetge, M. Krüger, and J. L. Velásquez, 'Algorithmic determination of pole-zero representations of power transformers' transfer functions for interpretation of FRA data', ISH 2009.

[40] S. A. Ryder, 'Transformer diagnosis using frequency response analysis: results from fault simulations', in *IEEE Power Engineering Society Summer Meeting,* 2002, pp. 399–404 vol.1. doi: 10.1109/PESS.2002.1043265.

[41] M. H. Samimi, S. Tenbohlen, A. A. S. Akmal, and H. Mohseni, 'Using the complex values of the frequency response to improve power transformer diagnostics', in *2016 24th Iranian Conference on Electrical Engineering (ICEE)*, 2016, pp. 1689–1693. doi: 10.1109/IranianCEE.2016.7585793.

[42] P. M. Nirgude, D. Ashokraju, A. D. Rajkumar, and B. P. Singh, 'Application of numerical evaluation techniques for interpreting frequency response measurements in power transformers', *IET Sci. Meas. Technol.*, vol. 2, no. 5, p. 275, 2008, doi: 10.1049/iet-smt:20070072.

[43] M. Wang, J. Vandermaar, and K. D. Srivastara, 'Evaluation of frequency response analysis data', *14th Int. Symp. High Volt. Eng.*, pp. 904–907, 2001.

[44] P. Karimifard, G. B. Gharehpetian, and S. Tenbohlen, 'Determination of axial displacement extent based on transformer winding transfer function estimation using vector-fitting method', *Eur. Trans. Electr. Power*, vol. 18, no. 4, pp. 423–436, 2008, doi: 10.1002/etep.194.

[45] DL/T 911-2016, *Frequency response analysis on winding deformation of power transformers*, 2016.

[46] G. M. Kennedy, A. J. McGrail, and J. A. Lapworth, 'Using Cross-Correlation Coefficients to Analyze Transformer Sweep Frequency Response Analysis (SFRA) Traces', in *2007 IEEE Power Engineering Society Conference and Exposition in Africa - PowerAfrica*, Johannesburg, South Africa: IEEE, Jul. 2007, pp. 1–6. doi: 10.1109/PESAFR.2007.4498059.

[47] J.-W. Kim, B. Park, S. C. Jeong, S. W. Kim, and P. Park, 'Fault diagnosis of a power transformer using an improved frequency-response analysis', *IEEE Trans. Power Deliv.*, vol. 20, no. 1, pp. 169–178, 2005, doi: 10.1109/TPWRD.2004.835428.

[48] K. P. Badgujar, M. Maoyafikuddin, and S. V. Kulkarni, 'Alternative statistical techniques for aiding SFRA diagnostics in transformers', *IET Gener. Transm. Distrib.*, vol. 6, no. 3, p. 189, 2012, doi: 10.1049/iet-gtd.2011.0268.

[49] W. C. Sant'Ana *et al.*, 'A survey on statistical indexes applied on frequency response analysis of electric machinery and a trend based approach for more reliable results', *Electr. Power Syst. Res.*, vol. 137, pp. 26–33, Aug. 2016, doi: 10.1016/j.epsr.2016.03.044.

[50] T. Y. Ji, W. H. Tang, and Q. H. Wu, 'Detection of power transformer winding deformation and variation of measurement connections using a hybrid winding model', *Electr. Power Syst. Res.*, vol. 87, pp. 39–46, 2012, doi: https://doi.org/10.1016/j.epsr.2012.01.007.

[51] V. Behjat and M. Mahvi, 'Statistical approach for interpretation of power transformers frequency response analysis results', *IET Sci. Meas. Technol.*, vol. 9, no. 3, pp. 367–375, May 2015, doi: 10.1049/iet-smt.2014.0097.

[52] W. H. Tang, A. Shintemirov, and Q. H. Wu, 'Detection of minor winding deformation fault in high frequency range for power transformer', in *IEEE PES General Meeting*, Minneapolis, MN: IEEE, Jul. 2010, pp. 1–6. doi: 10.1109/PES.2010.5589573.

[53] S. Miyazaki, M. Tahir, and S. Tenbohlen, 'Detection and quantitative diagnosis of axial displacement of transformer winding by frequency response analysis', *IET Gener. Transm. Distrib.*, vol. 13, no. 15, pp. 3493–3500, Aug. 2019, doi: 10.1049/iet-gtd.2018.6032.

[54] J. R. Secue and E. Mombello, 'Sweep frequency response analysis (SFRA) for the assessment of winding displacements and deformation in power transformers', *Electr. Power Syst. Res.*, vol. 78, no. 6, pp. 1119–1128, Jun. 2008, doi: 10.1016/j.epsr.2007.08.005.

[55] K. Pourhossein, G. B. Gharehpetian, E. Rahimpour, and B. N. Araabi, 'A probabilistic feature to determine type and extent of winding mechanical defects in power transformers', *Electr. Power Syst. Res.*, vol. 82, no. 1, pp. 1–10, Jan. 2012, doi: 10.1016/j.epsr.2011.08.010.

[56] M. H. Samimi, S. Tenbohlen, A. A. Shayegani Akmal, and H. Mohseni, 'Improving the numerical indices proposed for the FRA interpretation by including the phase response', *Int. J. Electr. Power Energy Syst.*, vol. 83, pp. 585–593, Dec. 2016, doi: 10.1016/j.ijepes.2016.04.044.

[57] E. Rahimpour and D. Gorzin, 'A new method for comparing the transfer function of transformers in order to detect the location and amount of winding faults', *Electr. Eng.*, vol. 88, no. 5, pp. 411–416, Jun. 2006, doi: 10.1007/s00202-005-0294-2.

[58] J. Bak-Jensen, B. Bak-Jensen, and S. D. Mikkelsen, 'Detection of faults and ageing phenomena in transformers by transfer functions', *IEEE Trans. Power Deliv.*, vol. 10, no. 1, pp. 308–314, 1995, doi: 10.1109/61.368384.

[59] M. Tahir, S. Tenbohlen, and M. H. Samimi, 'Evaluation of Numerical Indices for Objective Interpretation of Frequency Response to Detect Mechanical Faults in Power Transformers', in *Proceedings of the 21st International Symposium on High Voltage Engineering*, B. Németh, Ed., Cham: Springer International Publishing, 2020, pp. 811–824.

[60] L. Baccour and R. I. John, 'Experimental analysis of crisp similarity and distance measures', in *2014 6th International Conference of Soft Computing and Pattern Recognition (SoCPaR)*, 2014, pp. 96–100. doi: 10.1109/SOCPAR.2014.7007988.

[61] North China Electric Power Group Corp., NCEPRI 1999., *Application Guideline for Transformer Winding Distortion Test Technology*.

[62] S. Miyazaki *et al.*, 'Proposal of objective criterion in diagnosis of abnormalities of power-transformer winding by Frequency Response Analysis', in *2016 International Conference on Condition Monitoring and Diagnosis (CMD)*, Xi'an, China: IEEE, Sep. 2016, pp. 74–77. doi: 10.1109/CMD.2016.7757766.

[63] V. Larin and D. Matveev, 'Analysis of transformer frequency response deviations using whitebox modelling', in *CIGRE Study Committee A2 COLLOQUIUM, Cracow, Poland*, 2017.

[64] CIGRE JWG A2/C4.52, *High-Frequency Transformer And Reactor Models For Network Studies*. CIGRE Brochure 900;, 2023.

[65] B. Jurisic, I. Uglesic, A. Xemard, and F. Paladian, 'Difficulties in high frequency transformer modeling', *Electr. Power Syst. Res.*, vol. 138, pp. 25–32, Sep. 2016, doi: 10.1016/j.epsr.2016.02.009.

[66] CIGRE JWG A2/C4.52, *High-Frequency Transformer And Reactor Models For Network Studies*. CIGRE Brochure 901;, 2023.

[67] M. Hussain, N. Arbab, and A. Khan, 'High Frequency Modeling of Transformer Using Black Box Frequency Response Analysis', vol. 4, no. 10, Internation Journal of Eng 2017.

[68] CIGRE JWG A2/C4.52, *High-Frequency Transformer And Reactor Models For Network Studies*. CIGRE Brochure 902;, 2023.

[69] J. Pleite, E. Olias, A. Barrado, A. Lazaro, and J. Vazquez, 'Transformer modeling for FRA techniques', in *IEEE/PES Transmission and Distribution Conference and Exhibition*, 2002, pp. 317–321 vol.1. doi: 10.1109/TDC.2002.1178342.

[70] J. Pleite, E. Olias, A. Barrado, A. Lazaro, and R. Vazquez, 'Frequency response modeling for device analysis', in *IEEE 2002 28th Annual Conference of the Industrial Electronics Society. IECON 02*, 2002, pp. 1457–1462 vol.2. doi: 10.1109/IECON.2002.1185493.

[71] R. Aghmasheh, V. Rashtchi, and E. Rahimpour, 'Gray Box Modeling of Power Transformer Windings for Transient Studies', *IEEE Trans. Power Deliv.*, vol. 32, no. 5, pp. 2350–2359, 2017, doi: 10.1109/TPWRD.2017.2649484.

[72] E. Rahimpour, V. Rashtchi, and R. Aghmasheh, 'Parameters estimation of transformers gray box model', in *2017 International Conference on Modern Electrical and Energy Systems (MEES)*, Kremenchuk: IEEE, Nov. 2017, pp. 372–375. doi: 10.1109/MEES.2017.8248936.

[73] G. M. V. Zambrano, A. C. Ferreira, and L. P. Caloba, 'Power transformer equivalent circuit identification by artificial neural network using frequency response analysis', in *2006 IEEE Power Engineering Society General Meeting*, Montreal, Que., Canada: IEEE, 2006, p. 6 pp. doi: 10.1109/PES.2006.1708931.

[74] A. De and N. Chatterjee, 'Impulse fault diagnosis in power transformers using self-organising map and learning vector quantisation', *IEE Proc. - Gener. Transm. Distrib.*, vol. 148, no. 5, p. 397, 2001, doi: 10.1049/ip-gtd:20010462.

[75] S. Birlasekaran, Y. Xingzhou, F. Fetherstone, R. Abell, and R. Middleton, 'Diagnosis and identification of transformer faults from frequency response data', in *2000 IEEE Power Engineering Society Winter Meeting. Conference Proceedings (Cat. No.00CH37077)*, 2000, pp. 2251–2256 vol.3. doi: 10.1109/PESW.2000.847706.

[76] A. Akbari, P. Werle, H. Borsi, and E. Gockenbach, 'Transfer function-based partial discharge localization in power transformers: a feasibility study', *IEEE Electr. Insul. Mag.*, vol. 18, no. 5, pp. 22–32, 2002, doi: 10.1109/MEI.2002.1044318.

[77] S. Fei and X. Zhang, 'Fault diagnosis of power transformer based on support vector machine with genetic algorithm', *Expert Syst. Appl.*, vol. 36, no. 8, pp. 11352–11357, 2009, doi: https://doi.org/10.1016/j.eswa.2009.03.022.

[78] W. C. Flores, E. Mombello, J. A. Jardini, and G. Rattá, 'Fuzzy risk index for power transformer failures due to external short-circuits', *Electr. Power Syst. Res.*, vol. 79, no. 4, pp. 539–549, 2009, doi: https://doi.org/10.1016/j.epsr.2008.06.021.

[79] M. Tahir and S. Tenbohlen, 'Transformer Winding Condition Assessment Using Feedforward Artificial Neural Network and Frequency Response Measurements', *Energies*, vol. 14, no. 11, p. 3227, May 2021, doi: 10.3390/en14113227.

[80] *FRAnalyzer, Reliable core and winding diagnosis for power transformers - User's Manual.* [Online]. Available: https://www.omicronenergy.com/en/solution/sweep-frequency-response-sfra/

[81] *FRAX-series Sweep frequency analyser, User guide.* [Online]. Available: https://www.megger.com/en-gb/products/frax-series-sweep-frequency-response-analysers

[82] S. Miyazaki, Y. Mizutani, M. Tahir, and S. Tenbohlen, 'Influence of employing different measuring systems on measurement repeatability in frequency response analyses of power transformers', *IEEE Electr. Insul. Mag.*, vol. 35, no. 2, pp. 27–33, 2019, doi: 10.1109/MEI.2019.8636103.

[83] S. Miyazaki, Y. Mizutani, M. Tahir, and S. Tenbohlen, 'Comparison of FRA Data measured by Different Instruments with Different Frequency Resolution', in *2018 IEEE International Conference on High Voltage Engineering and Application (ICHVE)*, ATHENS, Greece: IEEE, Sep. 2018, pp. 1–4. doi: 10.1109/ICHVE.2018.8642033.

[84] M. Tahir, S. Tenbholen, and S. Miyazaki, 'Analysis of Statistical Methods for Assessment of Power Transformer Frequency Response Measurements', *IEEE Trans. Power Deliv.*, vol. 36, no. 2, pp. 618–626, Apr. 2021, doi: 10.1109/TPWRD.2020.2987205.

[85] IEC60076–5, 'Transformers - Part 5: Ability to withstand short circuit', *IEC Stand.*, 2006.

[86] S. Miyazaki *et al.*, 'Diagnosis Criterion of Abnormality of Transformer Winding by Frequency Response Analysis (FRA)', *Electr. Eng. Jpn.*, vol. 201, no. 3, pp. 25–34, Nov. 2017, doi: 10.1002/eej.23012.

[87] M. Tahir and S. Tenbohlen, 'Novel Calculation Method for Power Transformer Winding Fault Detection using Frequency Response Analysis'.

[88] T. Weiland, 'A discretization model for the solution of Maxwell's equations for six-component fields', *Arch. Elektron. Uebertragungstechnik*, vol. 31, pp. 116–120, 1977.

[89] *CST MICROWAVE STUDIO® 5 User's Manual.* [Online]. Available: www.cst.com

[90] S. Khaparde and S. Kulkarni, *Transformer Engineering: Design and Practice*. Taylor & Francis Limited, 2004.

[91] M. M. Tahir, S. Tenbohlen, and D. S. Miyazaki, 'Optimization of FRA by an Improved Numerical Winding Model: Disk Space Variation', VDE High Voltage Technology 2018.

[92] M. Tahir and S. Tenbohlen, 'FRA lookup charts for the quantitative determination of winding axial displacement fault in power transformers', *IET Electr. Power Appl.*, vol. 14, no. 12, pp. 2370–2377, Dec. 2020, doi: 10.1049/iet-epa.2020.0273.

[93] S. Tenbohlen, M. Tahir, E. Rahimpour, B. Poulin, and S. Miyazaki, 'A new approach for high frequency modelling of disc windings', in *CIGRE Paris*, 2018.

[94] M. Tahir and S. Tenbohlen, 'A Comprehensive Analysis of Windings Electrical and Mechanical Faults Using a High-Frequency Model', *Energies*, vol. 13, no. 1, p. 105, Dec. 2019, doi: 10.3390/en13010105.

[95] M. Heindl, S. Tenbohlen, J. Velásquez, A. Kraetge, and R. Wimmer, 'Transformer Modelling Based On Frequency Response Measurements For Winding Failure Detection', International Conference on Condition Monitoring and Diagnosis, Tokyo, Japan 2010.

[96] J. C. Gonzales and E. E. Mombello, 'Automatic detection of frequency ranges of power transformer transfer functions for evaluation by mathematical indicators', in *2012 Sixth IEEE/PES Transmission and Distribution: Latin America Conference and Exposition (T&D-LA)*, Montevideo: IEEE, Sep. 2012, pp. 1–8. doi: 10.1109/TDC-LA.2012.6319080.

[97] B. Cheng, P. Crossley, Z. Wang, P. Jarman, A. Fieldsend-Roxborough, and G. Wilson, 'Interpreting First Anti-resonance of FRA Responses Through Low Frequency Transformer Modelling', in *Proceedings of the 21st International Symposium on High Voltage Engineering*, vol. 598, B. Németh, Ed., in Lecture Notes in Electrical Engineering, vol. 598. , Cham: Springer International Publishing, 2020, pp. 982–992. doi: 10.1007/978-3-030-31676-1_92.

[98] R. S. (Great Britain) and R. S. (Great B. Staff, *Machine Learning: The Power and Promise of Computers That Learn by Example*. Royal Society, 2017. [Online]. Available: https://books.google.de/books?id=g-J0tAEACAAJ

[99] E. Alpaydin, *Introduction to Machine Learning, fourth edition*. in Adaptive Computation and Machine Learning series. MIT Press, 2020. [Online]. Available: https://books.google.de/books?id=uZnSDwAAQBAJ

[100] K. P. Murphy, *Machine Learning: A Probabilistic Perspective*. in Adaptive Computation and Machine Learning series. MIT Press, 2012. [Online]. Available: https://books.google.de/books?id=NZP6AQAAQBAJ

[101] S. Shalev-Shwartz and S. Ben-David, *Understanding Machine Learning: From Theory to Algorithms*. in Understanding Machine Learning: From Theory to Algorithms. Cambridge University Press, 2014. [Online]. Available: https://books.google.de/books?id=ttJkAwAAQBAJ

[102] D. P. Kumar, T. Amgoth, and C. S. R. Annavarapu, 'Machine learning algorithms for wireless sensor networks: A survey', *Inf. Fusion*, vol. 49, pp. 1–25, 2019, doi: https://doi.org/10.1016/j.inffus.2018.09.013.

[103] C. M. Bishop, *Pattern recognition and machine learning*. in Information science and statistics. New York: Springer, 2006.

[104] K. Grąbczewski, *Meta-Learning in Decision Tree Induction*. in Studies in Computational Intelligence. Springer International Publishing, 2013. [Online]. Available: https://books.google.de/books?id=TIO6BQAAQBAJ

[105] O. Z. Maimon and L. Rokach, *Data Mining With Decision Trees: Theory And Applications (2nd Edition)*. in Series In Machine Perception And Artificial Intelligence. World Scientific Publishing Company, 2014. [Online]. Available: https://books.google.de/books?id=OVYCCwAAQBAJ

[106] L. De Baets, C. Develder, T. Dhaene, and D. Deschrijver, 'Automated classification of appliances using elliptical fourier descriptors', in *2017 IEEE International Conference on Smart Grid Communications (SmartGridComm)*, 2017, pp. 153–158. doi: 10.1109/SmartGridComm.2017.8340669.

[107] F. Pedregosa *et al.*, 'Scikit-learn: Machine Learning in Python', *Mach. Learn. PYTHON*, Journal of Machine Learning Research 2011.

[108] M. Mohammed, M. B. Khan, and E. B. M. Bashier, *Machine Learning: Algorithms and Applications*. CRC Press, 2016. [Online]. Available: https://books.google.de/books?id=X8LBDAAAQBAJ

[109] I. H. Witten, E. Frank, and M. A. Hall, *Data Mining: Practical Machine Learning Tools and Techniques*. in The Morgan Kaufmann Series in Data Management Systems. Elsevier Science, 2011. [Online]. Available: https://books.google.de/books?id=bDtLM8CODsQC

[110] P. J. Braspenning, T. F, F. Thuijsman, and A. J. M. M. Weijters, *Artificial Neural Networks: An Introduction to ANN Theory and Practice*. in Lecture Notes in Computer Science. Springer, 1995. [Online]. Available: https://books.google.de/books?id=03DfRqUrTwsC

10 Appendix

A Liste of Publications

2023	-	M. Tahir and S. Tenbohlen, 'Transformer Winding Fault Classification and Condition Assessment Based on Random Forest Using FRA', Energies, vol. 16, no. 9, p. 3714, Apr. 2023.
2022	-	S. Tenbohlen et al., 'Use of Multiphysics Simulation Tools for Making a Digital Twin of Power Transformers', Cigré Paris Aug. 2022.
2021	-	M. Tahir and S. Tenbohlen, 'Transformer Winding Condition Assessment Using Feedforward Artificial Neural Network and Frequency Response Measurements', Energies, vol. 14, no. 11, p. 3227, May 2021.
	-	M. Tahir, S. Tenbholen, and S. Miyazaki, 'Analysis of Statistical Methods for Assessment of Power Transformer Frequency Response Measurements', IEEE Trans. Power Deliv., vol. 36, no. 2, pp. 618–626, Apr. 2021.
2020	-	M. Tahir and S. Tenbohlen, 'FRA lookup charts for the quantitative determination of winding axial displacement fault in power transformers', IET Electr. Power Appl., vol. 14, no. 12, pp. 2370–2377, Dec. 2020.
	-	M. Tahir, S. Tenbohlen, and M. H. Samimi, 'Evaluation of Numerical Indices for Objective Interpretation of Frequency Response to Detect Mechanical Faults in Power Transformers', in Proceedings of the 21st International Symposium on High Voltage Engineering, B. Németh, Ed., Cham: Springer International Publishing, 2020, pp. 811–824.
2019	-	M. Tahir and S. Tenbohlen, 'A Comprehensive Analysis of Windings Electrical and Mechanical Faults Using a High-Frequency Model', Energies, vol. 13, no. 1, p. 105, Dec. 2019.
	-	M. Tahir and S. Tenbohlen, 'Novel Calculation Method for Power Transformer Winding Fault Detection using Frequency Response Analysis', 5th International Colloquium, Opatija / Croatia 2019.
	-	S. Miyazaki, M. Tahir, and S. Tenbohlen, 'Detection and quantitative diagnosis of axial displacement of transformer winding by frequency response analysis', IET Gener. Transm. Distrib., vol. 13, no. 15, pp. 3493–3500, Aug. 2019.
	-	S. Miyazaki, Y. Mizutani, M. Tahir, and S. Tenbohlen, 'Influence of employing different measuring systems on measurement repeatability in frequency

		response analyses of power transformers', IEEE Electr. Insul. Mag., vol. 35, no. 2, pp. 27–33, 2019.
	-	M. Tahir and S. Tenbohlen, 'Transformer Equivalent Circuit Parameter Estimation using FRA', Cigré Colloquium SCA2; New Delhi, India, 2019.
2018	-	S. Miyazaki, M. Tahir, and S. Tenbohlen, 'Sensitivity of Connection Schemes for Detection of Axial Displacement of Transformer Winding by Frequency Response Analysis', in 2018 IEEE International Conference on High Voltage Engineering and Application (ICHVE), ATHENS, Greece: IEEE, Sep. 2018, pp. 1–4.
	-	S. Miyazaki, Y. Mizutani, M. Tahir, and S. Tenbohlen, 'Comparison of FRA Data measured by Different Instruments with Different Frequency Resolution', in 2018 IEEE International Conference on High Voltage Engineering and Application (ICHVE), ATHENS, Greece: IEEE, Sep. 2018, pp. 1–4.
	-	M. Tahir, S. Tenbohlen, and S. Miyazaki, 'Optimization of FRA by an Improved Numerical Winding Model: Disk Space Variation', VDE-Hochspannungstechnik, Berlin 2018.
	-	M. Tahir and S. Tenbohlen, 'Toward Objective Interpretation of Frequency Response of Power Transformers', ITCE 2018, Iran.
	-	M. Tahir, S. Tenbohlen, and S. Miyazaki, 'Development and Optimization of Methods for Objective Interpretation of Frequency Response Analysis of Power Transformers', 85th International Conference of Doble Clients, Boston 2018.
	-	S. Tenbohlen, M. Tahir, E. Rahimpour, B. Poulin, and S. Miyazaki, 'A New Approach for High-Frequency Modelling of Disc Windings', presented at the Cigré Paris 2018, pp. A2-214.

B Application of Adaptive Frequency Division Algorithm

The performance of the adaptive frequency division algorithm has been evaluated by different case studies. For this purpose, six case studies are presented where FRA measurements of different transformers having different type, rating, and voltage classes are employed. The results are presented in Figure B.1 – Figure B.4.

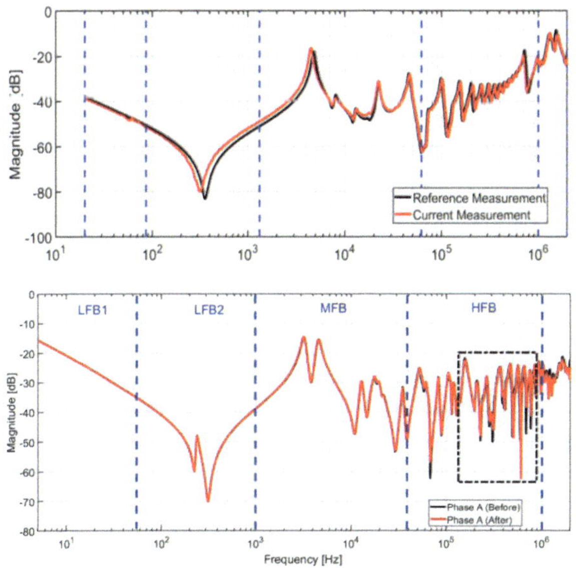

Figure B.1: Single-phase 170MVA, 735kV/315kV HV EE-OC (top) and 3-phase autotransformer, 240 MVA, 400/132 kV LV EE-OC phase A (bottom)

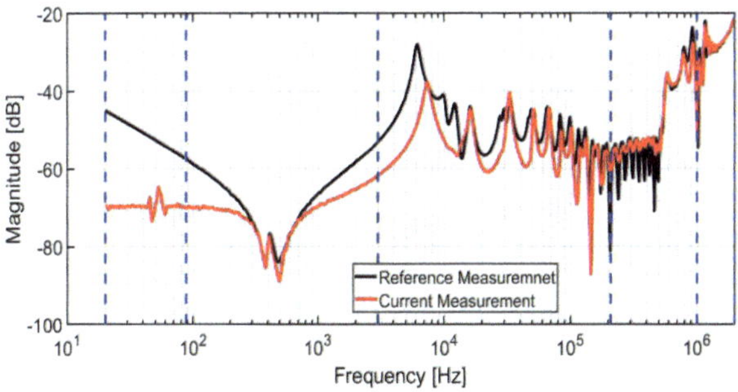

Figure B.2: 3-phase, 34 MVA, 237/5.65 kV transformer

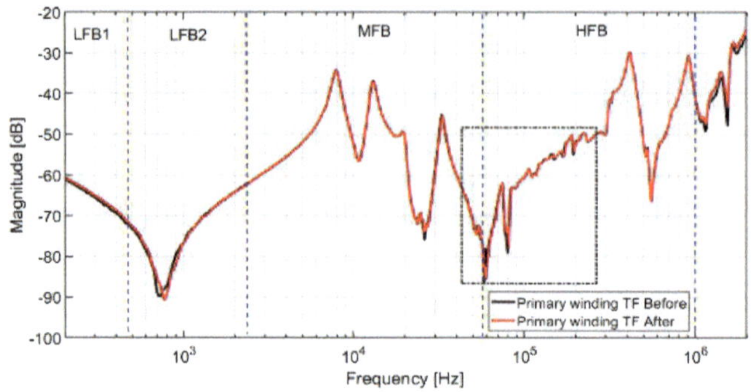

Figure B.3: 3-phase, 27 MVA, 154/10.5 kV transformer

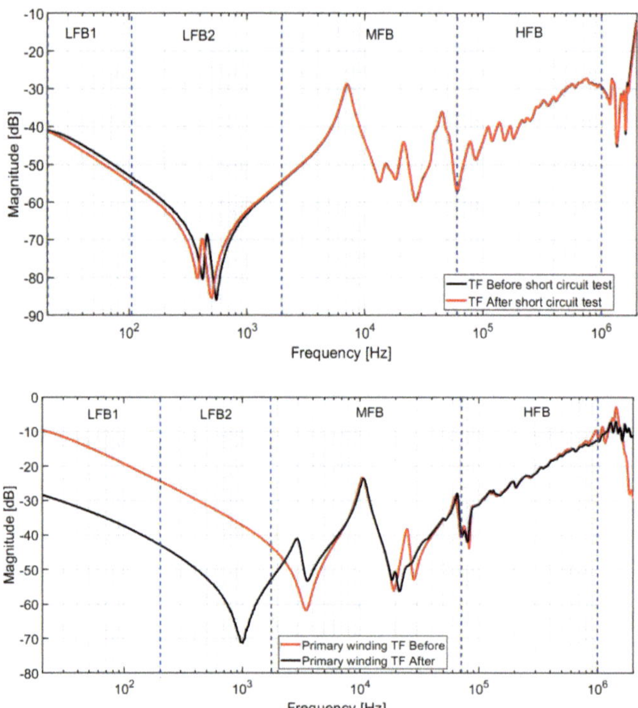

Figure B.4: 3-phase, 47 MVA, 120/26.4 kV, HV EE-OC phase A (top) and 3-phase 60 MVA, 105/6.6/22 kV, HV EE-OC phase B (bottom)

C Criteria of Monotonicity

The detailed evaluation results of monotonicity criteria for indices presented in Chapter 4 are presented in Table C.1.

Table C.1: Evaluation of index monotonicity criteria

Numerical Indices	Case 1				Case 2				Case 3				Case 4				Case 5				Case 6			
	OC	SC	CIW	IIW	OC	SC	CIW	IIW	OC	SC	CIW	IIW	OC	SC	CIW	IIW	OC	SC	CIW	IIW	OC	SC	CIW	IIW
Cross Correlation Factor (CCF)	✓	✓	✓	✓	✓	✓	✓	✓	✓	✓	✓	✓	✓	✓	✓	✓	✓	✓	✓	✓	✓	✓	✓	✓
Correlation Coefficient (CC)	✓	✓	✓	✓	✓	✓	✓	✓	✓	✓	✓	✓	✓	✓	✓	✓	✓	✓	✓	✓	✓	✓	✓	✓
Lin's Concordance Coefficient (CCC)	✓	✓	✓	✓	✓	✓	✓	✓	✓	✓	✓	✓	✓	✓	✓	✓	✓	✓	✓	✓	✓	✓	✓	✓
Complex Distance (CD)	✗	✗	✗	✗	✗	✓	✗	✗	✗	✓	✓	✓	✓	✓	✓	✓	✗	✓	✓	✓	✓	✓	✓	✓
Euclidean Distance (ED)	✓	✓	✓	✓	✓	✓	✓	✓	✓	✓	✓	✓	✓	✓	✓	✓	✓	✓	✓	✓	✓	✓	✓	✓
Standard Deviation (SD)	✓	✓	✓	✓	✓	✓	✓	✓	✓	✓	✓	✓	✓	✓	✓	✓	✓	✓	✓	✓	✓	✓	✓	✓
Comparative Standard Deviation (CSD)	✓	✓	✓	✓	✓	✓	✓	✓	✓	✓	✓	✓	✓	✓	✓	✓	✓	✓	✓	✓	✓	✓	✓	✓
Sum Squared Ratio Error (SSRE)	✓	✓	✓	✓	✓	✗	✓	✓	✓	✗	✓	✓	✓	✗	✓	✓	✓	✗	✓	✓	✓	✗	✓	✓
Sum Squared Error (SSE)	✓	✓	✓	✓	✓	✓	✓	✓	✓	✓	✓	✓	✓	✓	✓	✓	✓	✓	✓	✓	✓	✓	✓	✓
Maximum of Difference (MAD)	✗	✓	✓	✓	✗	✗	✓	✓	✓	✗	✗	✓	✓	✗	✗	✓	✓	✓	✗	✗	✓	✗	✗	✗
Stochastic Spectrum Deviation (SSD)	✓	✓	✓	✓	✓	✗	✓	✗	✓	✗	✓	✓	✓	✗	✓	✓	✓	✗	✓	✓	✓	✓	✓	✓
Expectation (E)	✓	✗	✓	✓	✓	✗	✗	✓	✓	✗	✗	✗	✓	✗	✓	✗	✓	✓	✓	✓	✓	✓	✓	✗
Root Mean square Error (RMSE)	✓	✓	✓	✓	✓	✓	✓	✓	✓	✓	✓	✓	✓	✓	✓	✓	✓	✓	✓	✓	✓	✓	✓	✓
Absolute Logarithmic Sum Error (ALSE)	✓	✓	✓	✓	✓	✓	✓	✓	✓	✓	✓	✓	✓	✓	✓	✓	✓	✓	✓	✓	✓	✓	✓	✓
Standardized Difference Area (SDA)	✓	✓	✓	✓	✓	✓	✓	✓	✓	✓	✓	✓	✓	✓	✓	✓	✓	✓	✓	✓	✓	✓	✓	✓
Absolute Integra Distance (AID)	✓	✓	✓	✓	✓	✓	✗	✓	✓	✓	✓	✓	✓	✓	✓	✓	✓	✓	✓	✓	✓	✓	✓	✓
Integra Distance (ID)	✓	✓	✓	✓	✗	✗	✓	✗	✗	✗	✗	✗	✓	✓	✓	✓	✓	✓	✓	✓	✓	✓	✓	✗
Minimum Maximum (MM)	✓	✓	✓	✓	✓	✓	✓	✓	✓	✓	✓	✓	✓	✓	✓	✓	✓	✓	✓	✓	✓	✓	✓	✓
Sum Squared Min Max Ratio (SSMMRE)	✓	✓	✓	✓	✓	✗	✗	✗	✓	✗	✗	✗	✗	✗	✗	✗	✓	✗	✗	✓	✓	✗	✗	✓
Sum of Error (SE)	✓	✓	✓	✓	✓	✓	✓	✓	✓	✓	✓	✓	✓	✓	✓	✓	✓	✓	✓	✓	✓	✓	✓	✓
Jaccard Distance (JD)	✓	✓	✓	✓	✓	✓	✓	✓	✓	✓	✓	✓	✓	✓	✓	✓	✓	✓	✓	✓	✓	✓	✓	✓
Least Squared Error (LSE)	✗	✓	✓	✓	✗	✗	✓	✗	✓	✗	✗	✗	✓	✗	✓	✓	✓	✗	✓	✗	✓	✓	✓	✓

D Geometrical details of Transformer model

The geometrical details of the transformer in Figure 5-2 are given in Table D.1.

Table D.1: Geometrical dimensions of the three-phase transformer

Parameters	Details
LV winding	Layer winding: 24 turns, 12 parallel conductors per turn
HV winding	Disc winding: 60 discs, 11 turns in each disc
Radius of core	10.5 cm
Inner radius of LV	11.5 cm
Inner radius of HV	14.57 cm
Height of core	134 cm
Windings height	85 cm
Tank height	136 cm
Tank width	52.8 cm
Tank length	132 cm

E Effect of winding structure on FRA

To investigate the effect of winding structure on FRA signatures. Mainly, two types of windings, i.e., continuous winding and interleaved winding, were simulated in the CST microwave studio. The FRA results are presented in Figure C-1. It can be seen that at high frequency the response of the continuous winding possesses more resonances compared to the frequency response of the interleaved winding. This is mainly due to the fact that the continuous winding possesses lower series capacitance. Its uniform winding structure gives the FRA response the exact 'U' shape feature exhibited in Figure E.1.

Due to the high series capacitance of interleaved disc windings, their FRA responses at high frequency would exhibit less resonances of which most are at low frequencies [8]. This is also demonstrated clearly in Figure E.1. The FRA response begins with a linear decreasing trend determined by the leakage inductance, then anti-resonance followed by the increasing trend of magnitude caused by the dominant high winding series capacitance. Referring to Figure E.1, the frequency response up to 10 kHz is determined by the leakage inductance of the windings. It can be seen that the frequency response of interleaved and continuous winding is well-matched up to 10 kHz, indicating that both winding models possess similar values of leakage inductance. Above 10 kHz, the capacitances start to exhibit resonant coupling effects, thus, producing anti-resonance. Its interleaving structure gives the FRA response the exact 'V' shape feature exhibited in Figure E.1.

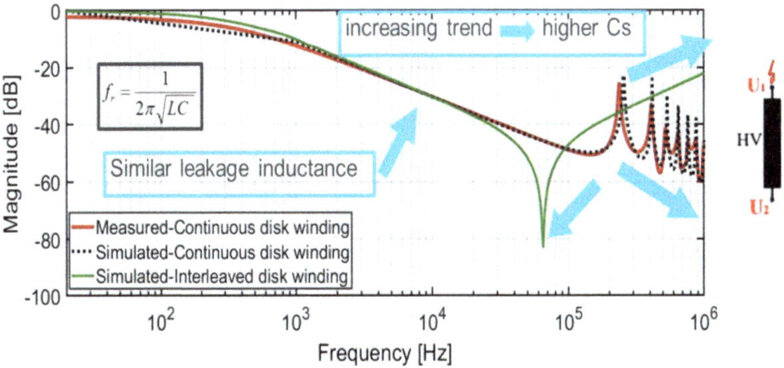

Figure E.1: FRA comparison of continuous and interleaved disk windings 60-disks

The characteristic trends of the frequency response of interleaved windings are also verified by the FRA measurements on HV and MV windings of a 300 MVA transformer during scraping as shown in Figure E.2. Where HV winding is 405 kV interleaved type disk winding and MV winding is 115 kV, a layer type winding. The measured FRA signatures are presented in Figure E.3[3]. As can be seen, the simulated FRA trace of the interleaved winding model (Figure E.1) and the FRA trace of the HV interleaved test winding shown in Figure E.3 look similar. Their interleaving structure produces the exact 'V' shape feature as shown in Figure E.1 and Figure E.3. Thus, verifying the accuracy of the simulation model of interleaved winding.

Figure E.2: FRA measurements at the winding of a 300 MVA transformer (a) 405 kV HV winding (interleaved disk winding) (b) 115 kV MV winding (layer winding)

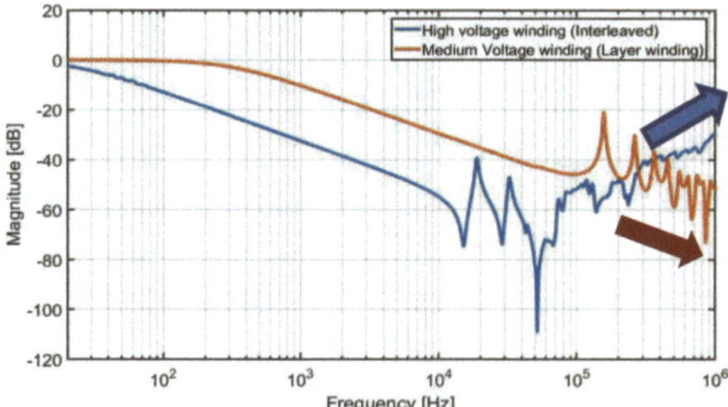

Figure E.3: FRA measurements of 405 kV HV winding (interleaved disk winding) and 115 kV MV winding (layer winding)

[3] Currently, due to the large computation space and time, it is not possible to model such large windings in the CST software.

F Fault simulation results – Interwinding connection scheme

This section includes the simulation results (Figure F.1 - Figure F.9) of interwinding connection schemes by simulating electrical and mechanical faults in HF model of transformer as presented in Chapter 5.

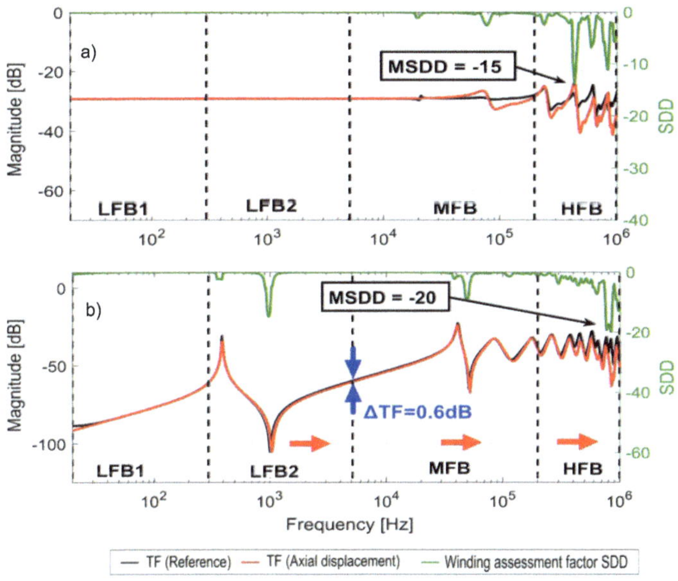

Figure F.1: TFs of healthy and axially displaced B-phase winding (vector group YNyn0), and SDD as a measure of deviation between TFs; (a) IIW (b) CIW

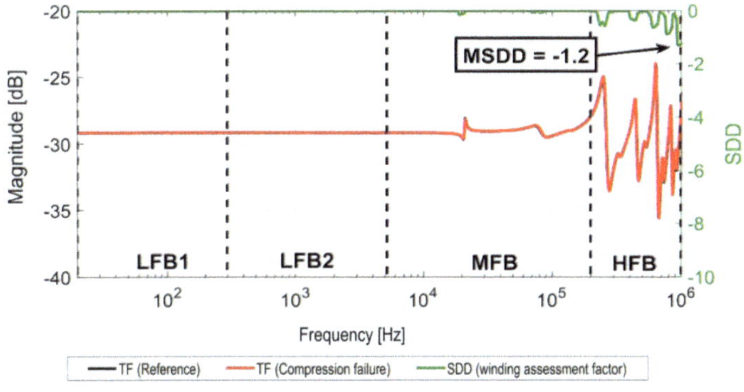

Figure F.2: CIW TFs of healthy and deformed B-phase LV winding (vector group YNyn0), SDD as a measure of deviation between TFs

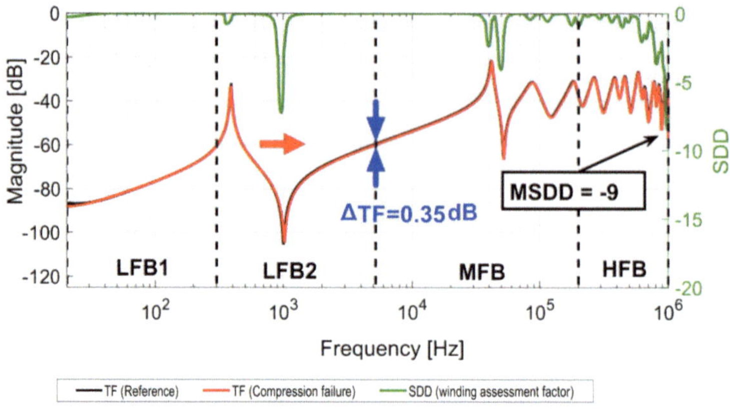

Figure F.3: IIW TFs of healthy and deformed B-phase LV winding (vector group YNyn0), SDD as a measure of deviation between TFs

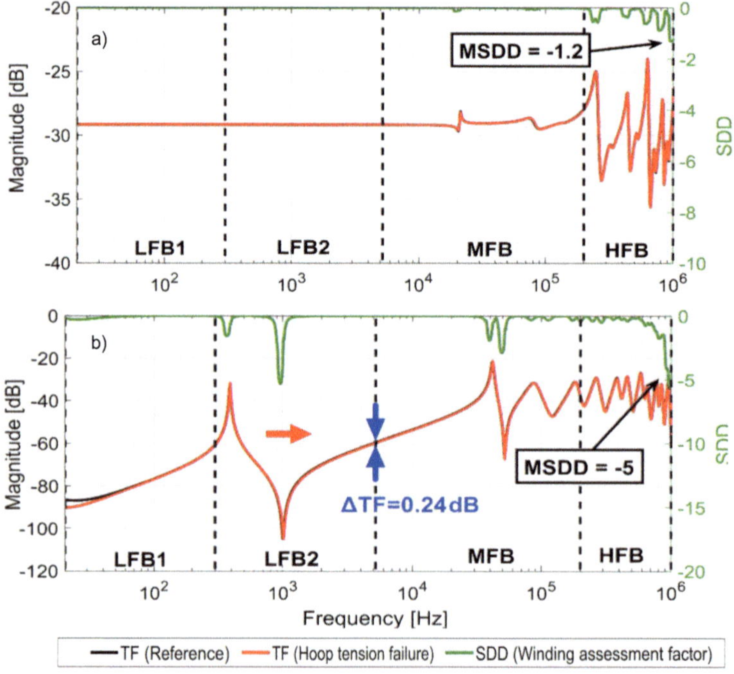

Figure F.4: TFs of healthy and deformed B-phase HV winding (vector group YNyn0) and SDD as a measure of deviation between TFs; (a) IIW (b) CIW

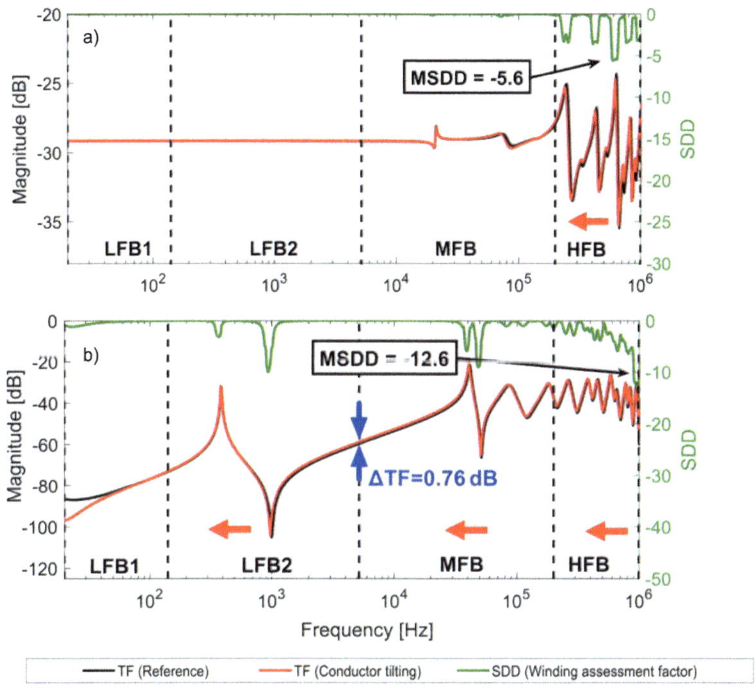

Figure F.5: TFs of healthy and deformed (conductor tilting failure) B-phase HV winding and SDD as a measure of deviation between TFs; (a) IIW (b) CIW

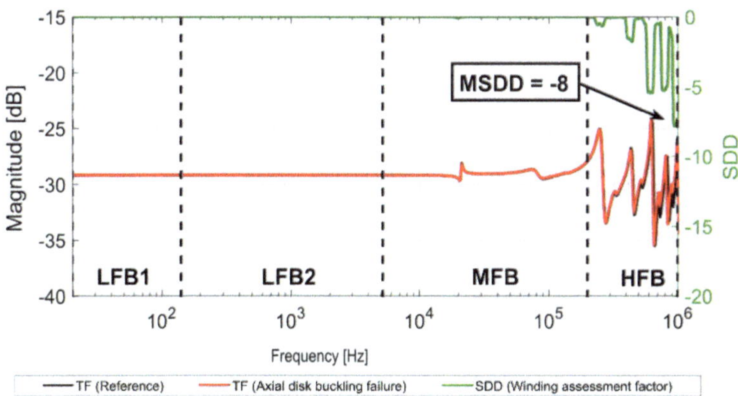

Figure F.6: IIW TFs of healthy and disk buckling failure in B-phase HV winding and SDD as a measure of deviation between TFs

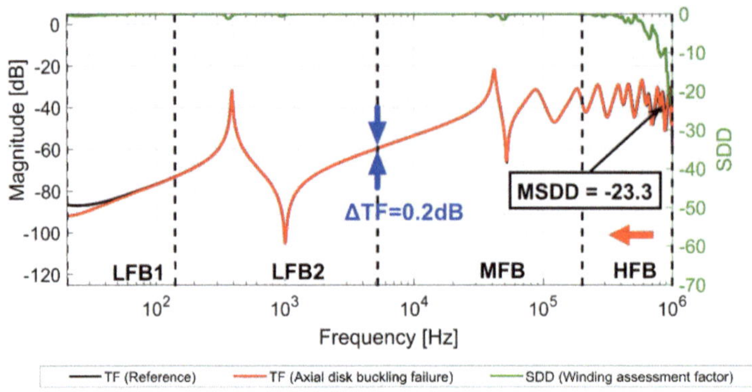

Figure F.7: CIW TFs of healthy and disk buckling failure in B-phase HV winding and SDD as a measure of deviation between TFs

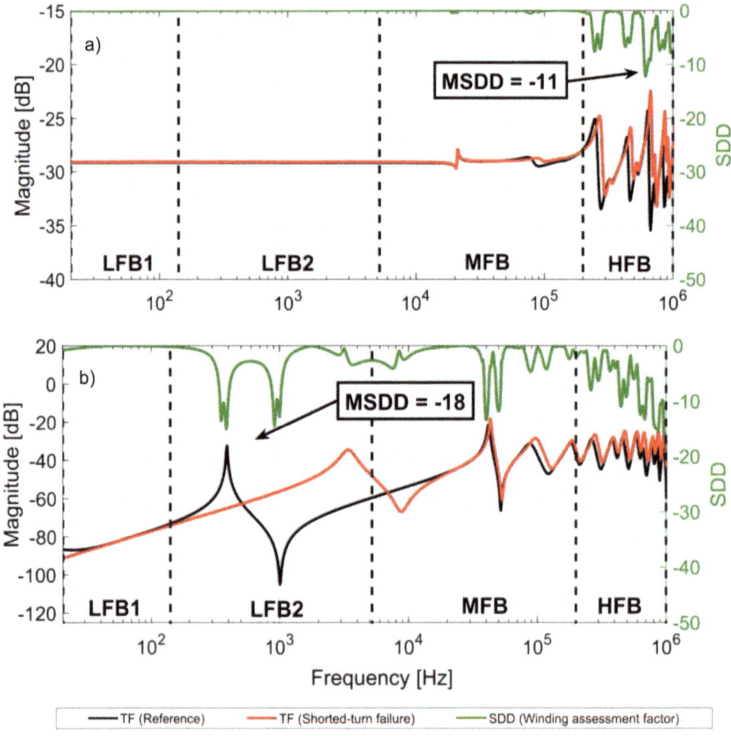

Figure F.8: TFs of healthy and SCC-E failure in B-phase HV winding and SDD as a measure of deviation between TFs; (a) IIW (b) CIW

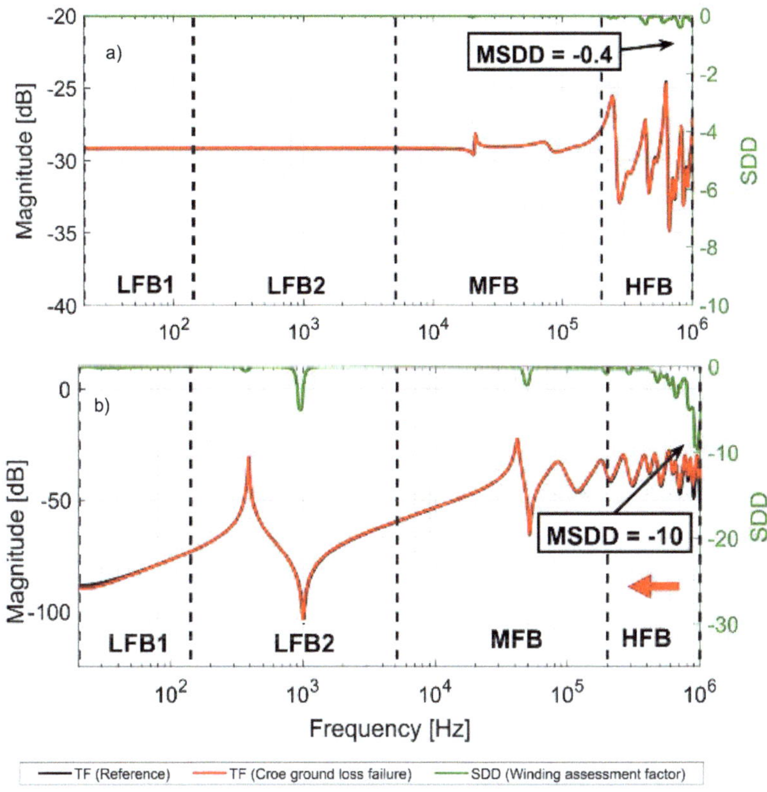

Figure F.9: TFs of B-phase HV winding with and without CGL-E (vector group YNyn0) and SDD as a measure of deviation between TFs; (a) IIW (b) CIW

G Theoretical Background of Machine Learning

G.1 Types of machine learning

Machine learning can be divided into three main types as shown in Figure G.1. A brief description of each type is given here.

G.1.1 Supervised learning

In supervised or predictive learning, the target is to infer a function or mapping from input x to output y, given a labeled set of training data. The training set D consists of N example input-output pairs $D = \{(x_i, y_i)\}_{i=1}^{N}$. The input vector x_i represents the features or attributes of data and output/response y_i is a discrete or continuous variable from some finite set $y_i \in \{1, \ldots, C\}$, where C represents the number of classes [100]. Such learning is called supervised learning because the output vector y contains labels for each training example. These labels are provided by the supervisor. These supervisors could be humans or machines. When y_i is continuous (real-valued) the problem is known as regression. In regression, models are built to predict continuous variables, such as using a curve to fit discrete data points including algorithms such as linear regression, polynomial regression, and stepwise regression [101]. When y_i is a discrete variable, the problem is known as classification or pattern recognition. If $C = 2$ ($y \in \{0, 1\}$), this is called binary classification, otherwise, if $C > 2$ this is called multi-class classification. The classification models are established to judge the category of the input data, using algorithms such as Decision Trees (DT), Artificial Neural Networks (ANNs), Support Vector Machines (SVM), and Naive Bayes (NB) [102].

G.1.2 Un-supervised learning

The second main type of ML is descriptive or unsupervised learning. Unsupervised learning does not require annotation of data and only input $D = \{(x_i)\}_{i=1}^{N}$ is given. The goal is to find interesting characteristics or patterns in the data. This approach is sometimes called knowledge discovery. Unsupervised learning is a much less well-defined problem because target patterns are not pre-defined and there is no obvious error metric to use [100]. Unsupervised learning is divided into clustering and dimensionality reduction. In clustering, data is clustered into different groups according to similarity, using algorithms such as hierarchical clustering, k-means, and fuzzy c-means. In the dimensionality reduction, the data is compressed while retaining the data structure and usefulness, using algorithms such as singular value decomposition (SVD), principal component analysis (PCA), and independent components analysis (ICA) [102].

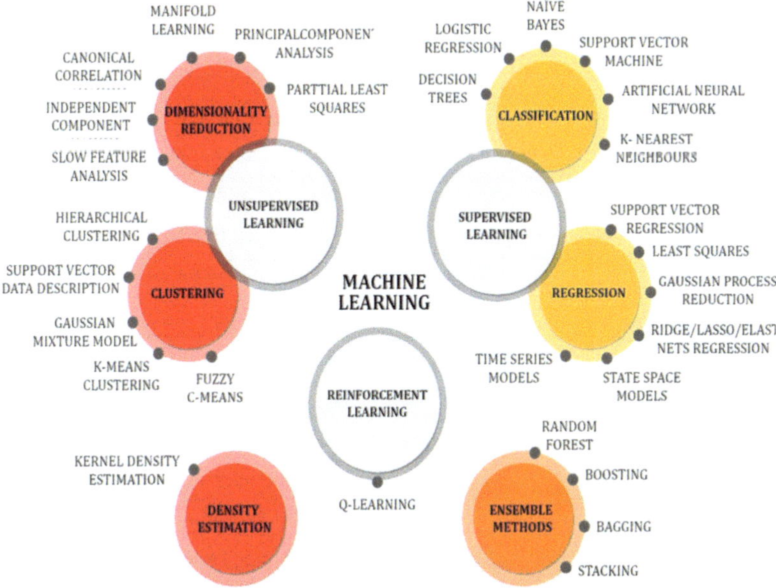

Figure G.1: Machine learning family tree

G.1.3 Reinforcement learning

Reinforcement learning lies between unsupervised and supervised learning and focuses on learning from experience. In this approach, an agent (the learning algorithm) uses observations gathered from the interaction with the environment to take actions that would maximize the reward or minimize the risk. The goal of the agent is to learn the consequences of its decisions and to use this learning to find strategies that maximize its rewards (a process of trial and error) [98], [103]. Reinforcement learning includes algorithms Q-learning, state-action-reward-state-action, deep Q network, and deep deterministic policy gradient [99].

G.2 Classification models of supervised learning

In the framework of this thesis six machine learning classification models namely Artificial neural network (ANN), decision tree (DT), random forest (RF), Naïve Bayes (NB), logistic regression (LR), and support vector machines (SVM) are studied for transformer winding condition assessment.

G.2.1 Decision Tree (DT)

A decision tree is a non-parametric predictive model that uses a tree-like structure or model of decisions to implement a divide-and-conquer strategy [99]. Decision tree algorithms use recursive portioning of data to achieve labels for the target variable. Decision trees are used for both classification and regression. The family of DT algorithms is usually called top-down induction on decision trees (TDIDT). It includes algorithms such as classification and regression trees (CART), concept learning systems (CLS), ID3 and C4.5, etc. [104].

A decision tree consists of a root node (RN), internal nodes (IN) also called decision nodes, leaf nodes (LN), or terminal nodes, branches, and splitters as shown in Figure G.2. Any internal node can be the root node of a sub-tree leading to the recursive nature of the decision tree. The leaf node contains instances belonging to a single class (pure node). The tree is generated by the recursive division of the training data set into smaller subsets using properly selected splitters (attributes) at each node until sufficiently pure nodes are obtained.

In general, the decision tree follows the steps outlined below:

- It puts all training data into a root node.
- It selects attributes by using some statistical measures.
- It divides the training dataset into subsets based on selected attributes.
- Recursive partitioning continues until no training example remains, or until no attribute remains, or all the remaining training data belongs to the same class.

As DT works on the divide-and-conquer strategy, the key for the division is based on the proper selection of splitter at each split. To determine the best features as splitters, mainly three measures are commonly used, i.e., Gini index (GI), information gain ratio (IGR), and information gain (IG). In this research, the C4.5 algorithm is employed, which is based on the information gain ratio.

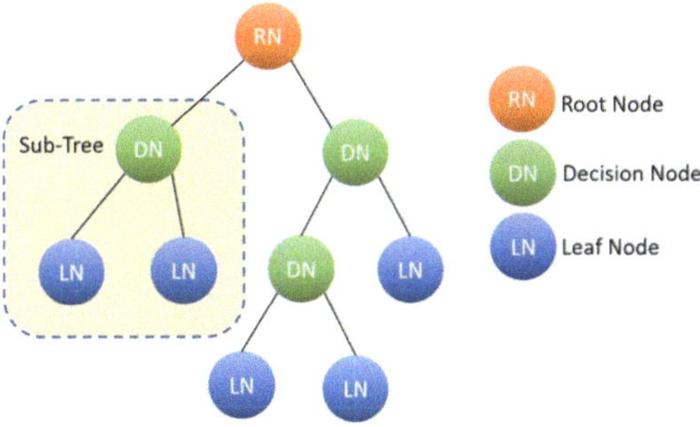

Figure G.2: Flow chart representation of decision tree

G.2.1.1 Information Gain Ratio (IGR)

The information gain ratio (IGR) is used to identify the features that give maximum information about a class. IGR is derived from information gain, which measures entropy reduction at each split. If the proportion of samples of class k in dataset D containing C classes is pk, the information entropy of D is defined as follows:

$$H(D) = -\sum_{k=1}^{C} p_k log_2 p_k \tag{10-1}$$

An attribute that provides a maximum reduction in entropy is selected as a splitter at the current node. Suppose the attribute x with N possible outcomes $x_1, x_2, \ldots x_n$ are used to divide the set D, N nodes are generated, wherein the nth node contains all samples in D that take the value of x_n on attribute x, denoted as D_n. The expected entropy if x is used as the current root is measured as

$$H_x(D) = \sum_{n=1}^{N} \frac{|D^n|}{|D|} H(D^n) \tag{10-2}$$

The information gained by selecting x to partition data is

$$Gain(D, x) = H(D) - H_x(D) \tag{10-3}$$

However, the information gain is biased toward the attribute with many distinct values. To overcome this problem, attributes that cause many splits are penalized.

Hence, the extent of partitioning is calculated by Splitinfo. This normalizes the information gain, resulting in the calculation of the gain ratio [105].

$$Splitinfo = -\sum_{n=1}^{N} \frac{|D^n|}{|D|} . \log_2 \frac{|D^n|}{|D|} \tag{10-4}$$

$$Gain\ ratio = \frac{Gain}{Splitinfo} \tag{10-5}$$

Pros

- Decision trees are easy to understand and interpret.
- Can handle both continuous and discrete data.
- Non-parametric model: no assumptions about the shape of data.
- Able to handle multi-output problems.
- Trees can be visualized.
- Can handle missing values in the dataset.

Cons

- Decision trees are prone to over-fitting
- Small variations in the data can change the tree shape completely and make the learning process unstable
- Decision tree learners are biased towards majority classes

G.2.2 Random Forest (RF)

Random forests are ensemble algorithms built on decision trees. An ensemble technique combines the predictions from multiple algorithms to make more accurate predictions than any individual model. In random forests, several trees are trained by randomly chosen subsets of the dataset that have the same size as the original training set. It then aggregates the votes from different decision trees to decide the final class of the test object [106]. Random forests improve predictive accuracy and control high variance and over-fitting in decision trees [107]. The training algorithm for random forests applies the bagging technique to tree learners. In bagging (bootstrap-aggregating), several subsets of data from the training sample are chosen randomly with replacement [100]. These new sets are called bootstrapped datasets. Individual trees are trained on the bootstrapped dataset.

Each tree predicts a class and the class with majority voting is assigned to the observation. A random forest model with *N* decision trees is shown in Figure G.3.

Pros

- Handles higher dimensionality data very well.
- Provides effective methods for estimating missing data.
- Highly stable and provides less variance than a single decision tree.

Cons

- Random forest does not provide precise prediction of continuous targets, hence, not suitable for regression.
- Acts like a black box model, provides little control over performance, and is less interpretable than an individual decision tree.
- Requires more computational resources.

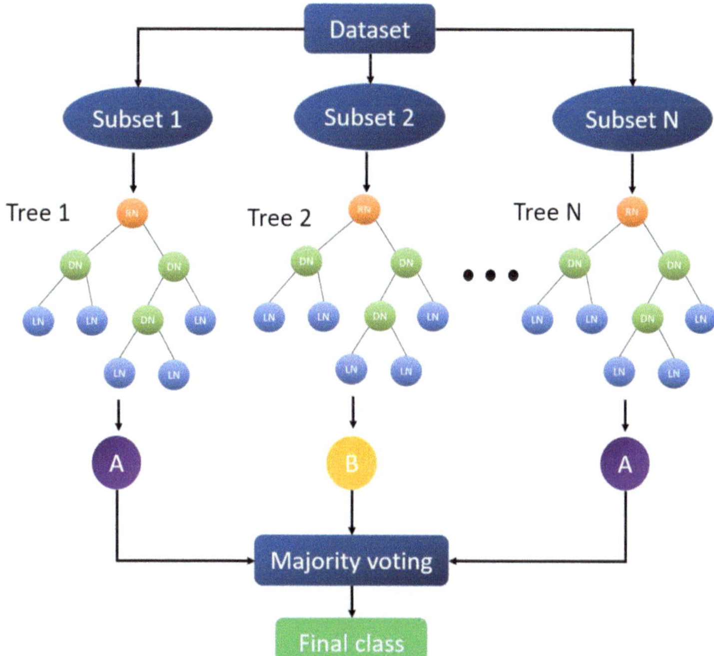

Figure G.3: Flow chart representation of random forest

G.2.3 Naive Bayes (NB)

In ML, Naive Bayes classifiers are a family of probabilistic algorithms that take advantage of probability theory and Bayes' theorem (with naive assumption) to predict the class of data. The fundamental Naive Bayes assumption is that features/attributes are conditionally independent and make an equal contribution to the output [100]. The assumptions made by Naive Bayes are generally incorrect in real-world situations. However, it performs surprisingly well even on datasets with substantial feature dependencies and often outperforms more sophisticated classification methods. Bayes theorem provides a way of calculating a posterior probability given a class variable C and feature vector $x = \{x_1, x_2 \ldots\ldots x_n\}$.

$$p(C|x) = \frac{p(x|C)p(C)}{p(x)} \tag{10-6}$$

Where,

$p(C|x)$ is the posterior probability of class (target) given predictors (attributes).

$p(C)$ is the prior probability of class.

$p(x|C)$ is the likelihood which is the probability of predictors given class.

$p(x)$ is the prior probability of attributes.

Being naive and assuming conditional independence, this relation is simplified to

$$p(C|x_1, x_2, \ldots x_n) = \frac{p(C)\prod_{i=1}^{n} p(x_i|C)}{\prod_{i=1}^{n} p(x_i)} \tag{10-7}$$

The denominator is independent of the C and plays no role in classification. Therefore, the denominator can be removed and proportionality can be introduced.

$$p(C|x_1, x_2, \ldots x_n) \propto p(C)\prod_{i=1}^{n} p(x_i|C) \tag{10-8}$$

The discussion so far has derived the independent feature model or the Naive Bayes probability model. The Naive Bayes classifier combines this model with a decision rule. One common rule is to find the probability of the given set of inputs for all possible values of the class variable C and pick up the output with maximum probability. This is called a maximum posteriori (MAP) decision rule.

$$\hat{C} = \underset{C}{\operatorname{argmax}}\{p(C)\prod_{i=1}^{n} p(x_i|C)\} \tag{10-9}$$

Despite their over-simplified assumptions, Naive Bayes classifiers have worked quite well in many real-world situations and require a small amount of training data to estimate the necessary parameters [107].

Pros

- Naive Bayes classifiers can be extremely fast compared to more sophisticated methods.
- Each class can be independently estimated as a one-dimensional distribution.
- Eliminates the curse of dimensionality.

Cons

- Suffers from "Zero Frequency" problem.
- Naive Bayes is known to be a bad estimator.
- The assumption of independence between features is often not true in real-world problems.

G.2.4 Support Vector Machine (SVM)

A support vector machine is a supervised machine learning algorithm used for classification and regression. SVM was first proposed for binary classification, however, it can be used for multi-class classification employing multiple binary classifiers. SVM builds a hyperplane or set of hyperplanes to classify all inputs in N-dimensional space (N=number of features). The instances closest to the classification margin are known as support vectors. The best hyperplane/decision boundary for an SVM is the one with the largest margin (minimum distance) between the two classes [108]. The dimension of the hyperplane depends upon the number of features. If the number of input features is 2, then the hyperplane is just a line. If the number of input features is 3, then the hyperplane becomes a two-dimensional plane. Figure G.4 shows a linear SVM with two features.

SVM is a generalized linear classifier that can draw a hyperplane between linearly separable data. However, in real-world problems where data is often non-separable SVM uses the kernel method or kernel trick. The kernel function maps the data onto a high-dimensional feature space and defines a separating hyperplane there.

A linear separation in feature space corresponds to a non-linear separation in the original input. The binary SVM classifier finds a hyperplane, which separates two groups of observations with known types. Given N training vectors $\{x_1, x_2 \ldots\ldots x_n\}$, with each vector x_i, belongs to one of two classes C_1 or C_2 the equation of the hyperplane is:

$$g(x) = w^T \cdot x + b = 0 \qquad (10\text{-}10)$$

Where x is the feature vector, w is the weighting vector and scalar b is biased. The weight vector is orthogonal to the hyperplane and controls its direction, whereas the bias controls its position. In the training stage of the SVM weight and bias are adjusted such that all the instances of C_1 lie on one side of the hyperplane, and the instances of C_2 lie on the other side of the hyperplane as shown in Figure G.4. SVM maximizes the margin (the distance between the hyperplane and the support vectors), such that instances from classes C_1 and C_2 are equally apart from the hyperplane. For a feature vector x_i:

$$g(x) = w^T \cdot x + b > 0 \quad \text{if } x_i \text{ is an instance in } C_1 \qquad (10\text{-}11)$$

$$g(x) = w^T \cdot x + b < 0 \quad \text{if } x_i \text{ is an instance in } C_2 \qquad (10\text{-}12)$$

Now for each $x_{i,}$ we choose y_i (classification label) such that:

$$y_i = f(x) = \begin{cases} 1, & x_i \epsilon C_1 \\ -1, & x_i \epsilon C_2 \end{cases} \qquad (10\text{-}13)$$

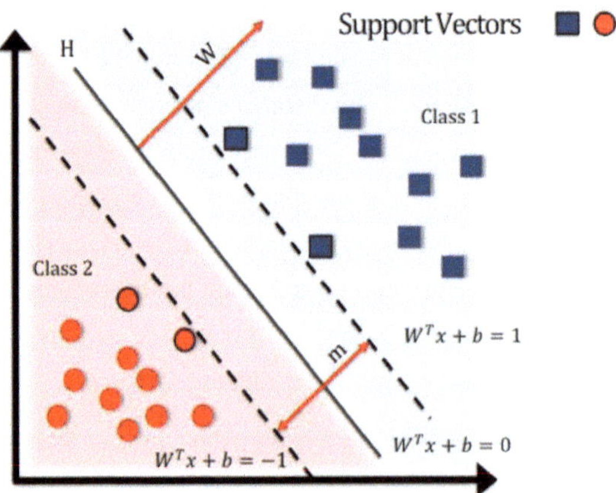

Figure G.4: SVM geometric definition

If x is any point in the feature space, then its distance from the hyperplane is given by:

$$z = \frac{|w^T \cdot x + b|}{||w||} \quad (10\text{-}14)$$

The numerator in (G-14) is equal to 1 if x is a support vector. The optimal hyperplane is achieved by minimizing the bias and maximizing the $||w||$, simultaneously. Once optimal w and b are obtained, the classification hyperplane is found. Therefore, the type of new features can be identified according to its sign.

The binary SVM can only classify two data classes. Hence, for multi-class classification, several binary SVM classifiers are used together. 'One-versus-one', and 'one-versus-all' are commonly adopted multiclass classification strategies.

Pros

- Effective in high-dimensional spaces.
- Performs well in cases where the number of dimensions is greater than the number of samples.
- Uses a small subset of training points in the decision function.
- Less complex model and memory efficient.
- Provides versatility through different Kernel functions.
- Model complexity is unaffected by the number of features.

Cons

- SVM does not directly provide probability estimates.
- Performance is poor if target classes are overlapping.
- SVM is not suitable for large datasets.
- Difficult to understand and interpret the final model, variable weights, and individual impact.
- Choice of best kernel might be tricky.

G.2.5 Logistic Regression (LR)

Logistic regression is a classification algorithm used to assign observations to a discrete set of classes depending on the feature values. Unlike linear regression where outputs are continuous values, logistic regression converts its output to a probability value using the logistic or sigmoid function. The output probability values can then be mapped to two or more discrete classes [109].

Considering binary classification with two classes C_1 and C_2, suppose we have an input observation vector x with N features $\{x_1, x_2.....x_n\}$. The classifier output y_i can be 1 (observation belongs to C_1) or 0 (the observation belongs to C_2). We want to know the probability $p(y = 1|x)$ and $p(y = 0|x)$ that this observation is a member of a particular class. In logistic regression, this task is solved by learning from a training set, a vector of weights, and a bias term. Each input feature x_i has weight w_i associated with it. The weight w_i represents the significance of that input feature in the classification decision and can be positive or negative. The bias or intercept, also a real number is added to the weighted inputs. After learning the weights in training, the classifier decides on the test instance by first multiplying each x_i by its weight w_i, sums up the weighted features, and adds the bias. The resulting number z represents the weighted sum of the evidence for the class.

$$z = \left(\sum_{i=1}^{n} w_i x_i\right) + b = w \cdot x + b \qquad (10\text{-}15)$$

z is also a real number between $-\infty$ and $+\infty$. To create probabilities z is transformed through the sigmoid function, $\sigma(z)$ as shown in Figure G.5.

$$y = \sigma(z) = \frac{1}{1 + e^{-z}} \qquad (10\text{-}16)$$

Finally, the probability that an observation belongs to a specific class or not is calculated.

$$p(y = 1|x) = \frac{1}{1 + e^{-z}} \qquad (10\text{-}17)$$

$$p(y = 0|x) = 1 - \frac{1}{1 + e^{-z}} = \frac{e^{-z}}{1 + e^{-z}} \qquad (10\text{-}18)$$

Deciding for a test instance x, if the probability $p(y = 1|x)$ is more than 0.5, it belongs to C_1 otherwise it belongs to C_2. Hence, 0.5 is called the threshold value or decision boundary.

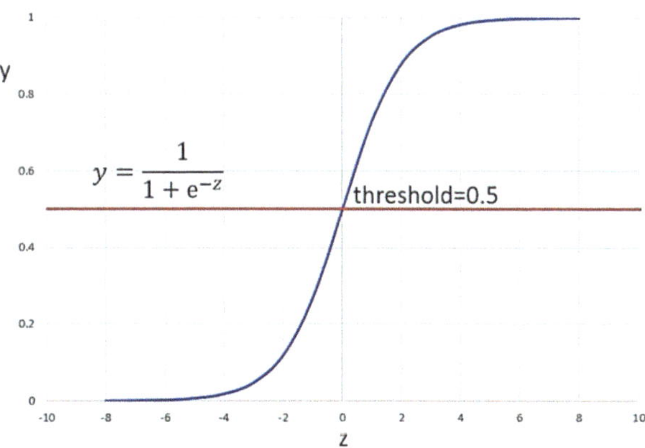

Figure G.5: Binary logistic function

Pros

- Logistic regression is very robust and easy to interpret.
- Provides coefficient size and direction of association for each predictor.

Cons

- Cannot solve non-linear problems.
- Does not perform well for data with a large number of features and classes.
- The assumption of indecency between dependent and independent variables is rarely true.
- May over-fit with a large number of features.

G.2.6 Artificial Neural Network (ANN)

The artificial neural network is an information processing model which is inspired by the biological neuron system. It is composed of a large number of highly interconnected processing elements known as neurons to solve problems [110]. The network can be trained or adjusted, so that a particular input leads to a specific output. Generally, the network is adjusted (weights and bias) by comparing the output and the target, until the network output matches the data target. Thus, a neural network is also an adaptive system. According to the recursive characteristic of a network, two main types of networks can be classified: feedforward and feedback artificial neural network. In a feedforward network, the neurons in one layer are only connected to neurons in the next layer, while they are connected to the neurons in both layers (last and next) in the feedback network. The feedback network can be applied for the data inputs which are changed over time. Therefore, a feedforward network is suitable for the classification of the six classes.

To select the optimum structure of ANN, a sensitivity study is performed. For this purpose, overall accuracy is compared using a different number of hidden layers, and five commonly used activation functions are tested in these hidden layers. Each hidden layer consists of eight neurons. Increasing the number of neurons in the hidden layer increases the power of the network but requires more computation and is more likely to produce overfitting. The neurons of the hidden layer are chosen through trial and error. Hidden neurons are increased if the training performance is poor, while they are decreased if the training performance shows overfitting.

In this work, the hidden layer with eight neurons gives a better training performance. The results of the sensitivity study are presented in Table G.1. It is clear that tansig and logsig outperformed the other activation functions. It is important to note that accuracy has minimal impact on increasing the number of hidden layers. However, the computation cost and complexity increase with the increase of hidden layers. This may lead to overfitting. To avoid these issues, a simple one hidden layer feedforward network with a tansig activation function for the hidden layer and a linear activation function for the output layer is used in this work. Figure G.6 demonstrates an associative ANN that consists of an input layer with 'r' nodes, a hidden layer with 'm' nodes, and an output layer with 'n' nodes. The neurons of the input layer receive four features as inputs from different indicators (139 × 4 matrix). The output vector is a column matrix with six rows that correspond to six classes (6 × 1 matrix). The signal from the input layer is transmitted to the output layer through the hidden layer. The relationship between input vector F_i, hidden layer vector h_j, and the output vector y_k can be represented by Equations (G-19) and (G-20):

$$h_j = f\left(\sum_{i=1}^{r} W_{ij} F_i + b_j\right) \tag{10-19}$$

$$y_k = f\left(\sum_{j=1}^{m} W_{jk} h_j + b_k\right) \tag{10-20}$$

$$i = 1,2,\ldots r,\ j = 1,2,\ldots m,\ k = 1,2,\ldots n, \tag{10-21}$$

Where F_i is the input vector, W_{ij} is the weight vector or connection strength between the input and hidden layer neurons, b_j is the bias of the hidden layer neuron, W_{jk} is the weight vector or connection strength of hidden and output layer neurons, and b_k is the bias of output layer neuron.

Pros

- ANNs are also able to tolerate faults, meaning that they can still function correctly even if some of the neurons in the network are damaged or destroyed.

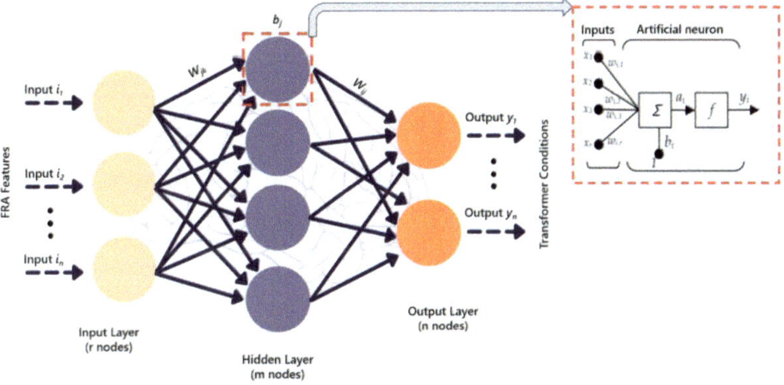

Figure G.6: Structure of feedforward artificial neural network

Table G.1: Percentage accuracy for different activation functions and number of hidden layers

Activation Function	**Abbr.**	**Percentage Accuracy in different number of Hidden Layers**				
		1	**2**	**3**	**4**	**5**
Linear	purelin	60.1	60.9	60.9	60.9	60.9
Triangular basis	tribas	92	94.2	94.2	95.7	94.8
Radial basis	radbas	94.2	96.4	94.9	91.3	94.9
Hyperbolic tangent sigmoid	tansig	97.1	96.6	97.1	96.6	96.6
Log-sigmoid	logsig	95.3	95.8	96.3	94.8	97.6

- ANN frameworks can take advantage of parallel and distributed computing, leading to faster training and inference times.
- ANNs can automatically learn relevant features from raw data, reducing the need for extensive manual feature engineering

Cons

- There is no specific rule for determining the structure of ANN. Appropriate network structure is achieved through experience and trial and error methods.
- ANNs require large amounts of data for training, and the quality and quantity of data significantly impact their performance.
- Designing, training, and tuning ANNs can be complex and resource-intensive.
- ANNs are prone to overfitting.

F.3 Performance Metrics

The performance metrics are used to evaluate a machine-learning algorithm to determine whether it is performing satisfactorily or not. The performance of a classifier can be measured by two ways, i.e., - general performance and detailed performance for each class. Both methods can be used simultaneously, especially when different classifiers are compared or combined.

G.3.1 General Performance Metrics

These metrics are used to measure the general performance of the classifier in categorizing all classes. However, they do not provide information about how well each particular class was classified. General performance metrics include accuracy, kappa static, mean square error, root mean square error, etc. In this work, among these metrics, only the accuracy is used and hence defined here.

G.3.1.1 Accuracy

Classification accuracy is the percentage of correctly classified instances. It is calculated by the ratio of correctly classified instances/samples and the total number of instances in the dataset.

$$Accuracy = \frac{\text{Number of correctly classified instances}}{\text{Total instances in the dataset}} \times 100 \qquad (10\text{-}22)$$

Accuracy is a very simple and intuitive measure that can immediately provide information about model performance. However, it could be misleading especially for imbalanced data where the number of instances in one class is much higher than the other. For example, if there are 90 faulted and 10 healthy transformers in our dataset, an accuracy of 90% can be obtained without predicting any healthy transformer, which makes it a poor measure if classes are imbalanced.

G.3.2 Detailed performance metrics (per class)

To evaluate and compare different classifiers, it is important to know the general performance of the classifier as well as a detailed performance per class (how good individual classes can be classified). It is important to note that a classifier with a poor general classification performance, might perform very well for specific classes. Thus, these metrics are important to assess the overall performance of a classifier. Some of the metrics used in this work are described below.

G.3.2.1 Confusion matrix (CM)

A confusion matrix also called the error matrix is a table that summarizes the predictive performance of a classification model on given labels. It contains the number of instances for each class that has been correctly or incorrectly classified. A confusion matrix is a good option to report results for both binary and multi-class classification problems. For the above two-class problem of faulted and healthy transformers, the confusion matrix is shown in Table G.2.

The diagonal contains correct predictions while incorrect predictions lie on off-diagonal cells.

Table G.2: Basic representation of a confusion matrix

		Predicted Class	
		Healthy	**Faulted**
Actual Class	**Healthy**	TP	FN
	Faulted	FP	TN

Definition of the Terms:

True Positive (TP): Transformer is healthy (positive), and is predicted to be healthy (positive).

False Negative (FN): Transformer is healthy (positive), but is predicted faulted (negative).

True Negative (TN): Transformer is faulted (negative), and is predicted to be faulted (negative).

False Positive (FP): Transformer is faulted (negative), but is predicted healthy (positive).

G.3.2.2 Precision

Precision measures how many instances classified as "positive" class are indeed "positive". Precision is calculated as the ratio of correctly predicted positive observations to the total predicted positive observations.

$$Precision = \frac{TP}{TP + FP} \tag{10-23}$$

Precision answers this question "of all the transformers that are predicted as healthy, how many are actually healthy".

G.3.2.3 Recall

Recall assesses how well the classifier can identify positive instances and is defined as the ratio of correctly predicted positive instances to all instances in the actual class.

$$Recall = \frac{TP}{TP + FN} \tag{10-24}$$

The question recall answers, "From all the transformers which are truly healthy, how many are identified". In ML recall is also called sensitivity, true positive rate, and hit rate.

G.3.2.4 F-score

Both precision and recall are important metrics to measure the performance of classifiers. However, maximizing both at the same time is often not possible. Usually, there is a trade-off between the two, and improving precision degrades the recall and vice versa. To convey a balance between precision and recall, F-score is used which is the harmonic mean between precision and recall. The greater the F-score, the better the performance of the classification model.

$$F - score = 2 \times \frac{precision \times recall}{precision \times recall} \tag{10-25}$$

H FRA Graphical User Interface

As a first attempt to develop an automatic assessment tool a MATLAB-based graphical user interface (GUI) is developed. This FRA application allows the user to import the raw FRA measurement data in Excel format and perform an automatic assessment of FRA results based on the winding assessment factor.

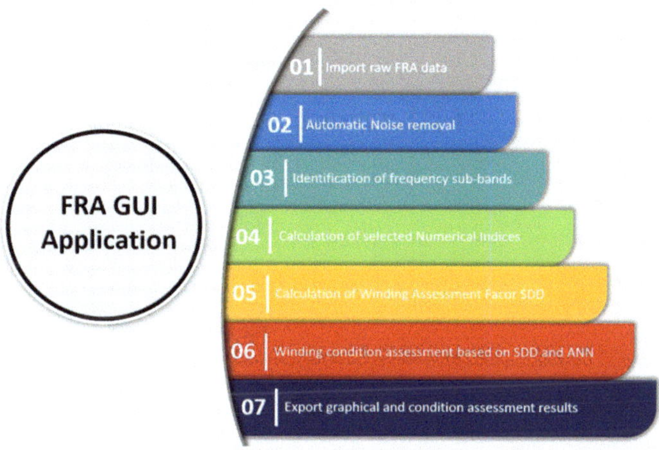

Figure H.1: The key features of the FRA GUI application

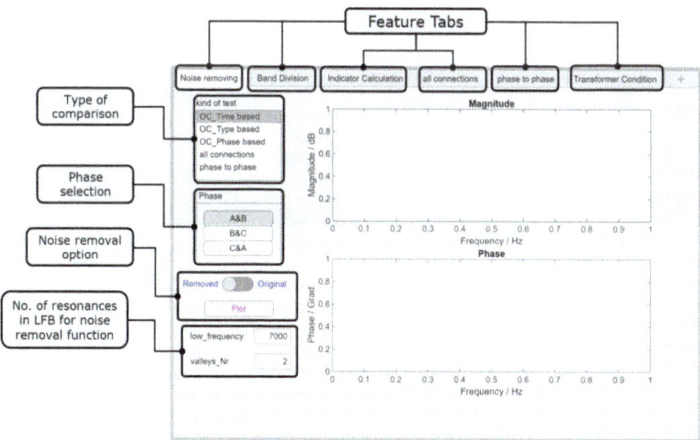

Figure H.2: Pictorial representation of FRA application

In addition to the index analysis, the algorithm of ANN is also integrated into the application. Thus, a two-way assessment is possible with the FRA application. The key features of the FRA application are stated in Figure H.1. A pictorial representation of the FRA application along with different features is shown in Figure H.2 – Figure H.5.

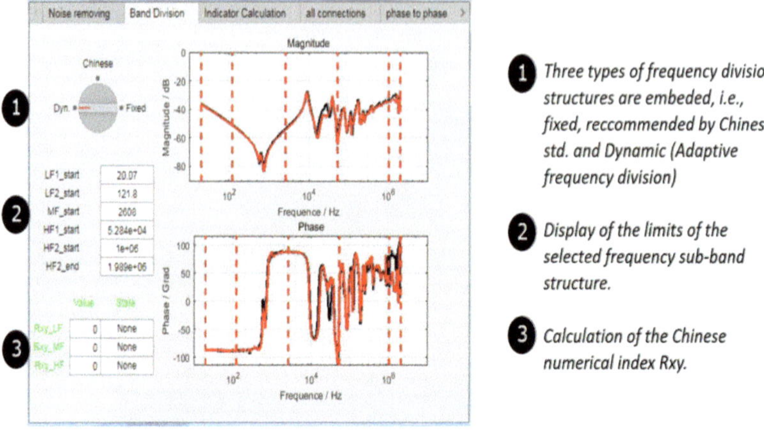

Figure H.3: Representation of frequency sub-band division feature in FRA application

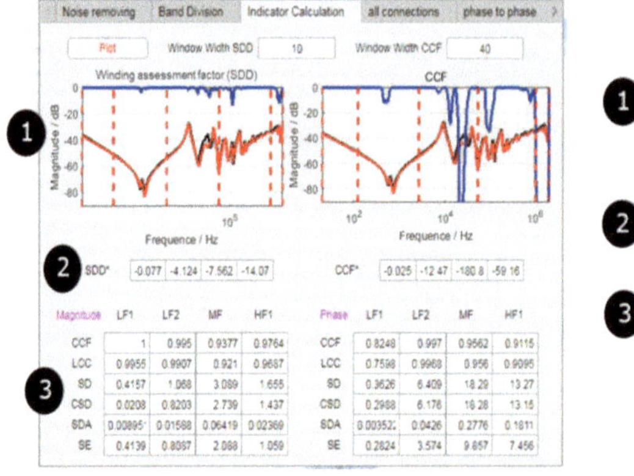

1 *Graphical representation of FRA comparison and application of winding assessment factor SDD.*

2 *Calculated values of SDD for each frequency sub-band*

3 *Calculation of slected numerical indices for each frequency sub-band*

Figure H.4: Representation of index calculation feature

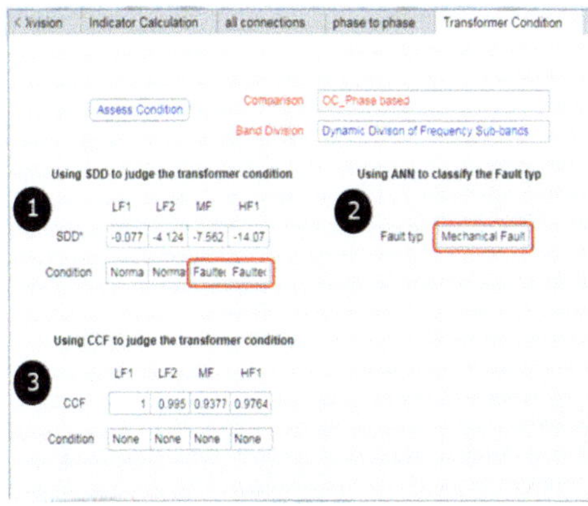

1 *Transformer winding conition assessment based on winding assessment factor SDD – based on proposed winding assessment criteria.*

2 *Condition assessment and type of fault identification based on Artificial Neural Network ANN*

3 *Condition assessment based on CCF - Doble criteria*

Figure H.5: Transformer winding condition assessment and fault identification based on winding assessment factor (SDD) and ANN